Michael J. Ruf

Sustainability and Evolut

German University Press (GUP)

An imprint of the *Deutscher Wissenschafts-Verlag (DWV) Baden-Baden*

Michael J. Ruf

Sustainability and Evolution

or why life becomes increasingly complex:

The *Interaction Theory*

German University Press (GUP)

An imprint of the *Deutscher Wissenschafts-Verlag (DWV) Baden-Baden*

Umschlaggestaltung
DWV in Zusammenarbeit mit dem Autor

Bibliografische Information Der Deutschen Nationalbibliothek
Die Deutsche Nationalbibliothek verzeichnet diese Publikation in der
Deutschen Nationalbibliografie; detaillierte bibliografische Daten sind
im Internet über http://dnb.dnb.de abrufbar.
**Bibliographic information published by Die Deutsche
Nationalbibliothek**
Die Deutsche Nationalbibliothek lists this publication in the Deutsche
Nationalbibliografie; detailed bibliographic data are available in the In-
ternet at http://dnb.dnb.de.
Information bibliographique de Die Deutsche Nationalbibliothek
Die Deutsche Nationalbibliothek a répertorié cette publication dans la
Deutsche Nationalbibliografie; les données bibliographiques détaillées
peuvent être consultées sur Internet à l'adresse http://dnb.dnb.de.

1. Auflage
Gedruckt auf alterungsbeständigem, chlorfrei gebleichtem Papier

© Copyright 2018 by
Deutscher Wissenschafts-Verlag (DWV)®
Postfach 11 01 35
D–76487 Baden-Baden

www.dwv-net.de
www.UniversityPress.de

ISBN: 978-3-86888-133-2

Table of Contents

Preface   3

Introduction   5

Part 1: The formulation of *Interaction Theory*   **14**

1.1. The *Interaction Theory*   14
1.2. The framework to study generation sequences   19
1.3. The fate of generation sequences   23
1.4. The niche requirements of generation sequences   29
1.5. The two fundamental types of niche   34
1.6. The role and significance of reproductive interactions   37
1.7. Symbiosis and generation sequences   50
1.8. Complexity increase and the probability of multiplication   54
1.9. A rising probability of multiplication   63
1.10. The way in which the probability of multiplication can rise   72
1.11. Environmental changes and complexity increase   86
1.12. The law-like process of complexity increase   100

Part 2: The course of complexity increase during evolution   **104**

2.1. The journey of generation sequences through deep time   104
2.2. From molecules to prokaryotes   107
2.3. The emergence of eukaryotes   117
2.4. The emergence of sexual reproduction   127
2.5. The evolution of sexually reproducing species   133
2.6. Species with social interaction   143
2.7. The evolution of *Homo sapiens*   148

Part 3: What is life?   **161**

Part 4: The consequences of *Interaction Theory* beyond biology   **164**

Flowchart 1   172

Annex   173

Notes   179

"The electric light did not come from the continuous improvement of candles."
(Oren Haran)

"The purpose of evolution, believe it or not, is beauty and not fitness."
(Based on Joseph Brodsky)

At first sight the theory of evolution looks simple and clear. The trouble is, however, that the interpretation of how evolution exactly works is inevitably influenced by the prevailing zeitgeist: with Darwin in the Victorian era it was an authoritarian natural selection that rules mercilessly in a world of teeth, claws and blood, whereas in the modern age of globalization with its fierce economic competition, it is the blind belief in fitness as the driving force of evolution. Today however, it is high time to focus on the obvious fact that life needs sustainability and an ecological balance between growth and exhaustible resources. With this, the purpose of evolution becomes indeed beauty and not fitness.

Preface

This book is written for all readers interested in the evolution of life, and who are familiar with the main principles of modern biology. Contrary to most other books about evolution and life, it is focused on a specific subject, the complexity increase that has occurred during evolution. For this particular and essential aspect of life, the book provides a radically new way of thinking, why and how increasingly complex biological forms have evolved. On the following pages I will therefore introduce the corresponding new theory with the name *Interaction* Theory. A first, preliminary version of *Interaction Theory* was published in 2003 in German (Ruf M. 2003).

When it comes to the evolution of life, it is nowadays better to avoid any misunderstanding. I am not a creationist, nor do I sympathize with this movement. On the contrary, I believe in science and its way to approach the world. This includes my respect for Darwin's theory of evolution as an important scientific achievement. With regard to biological complexity however, I deeply believe that the existing evolutionary theory can only be part of the story. And the same can be said with regard to molecular biology, which offers impressive insights about the functioning of the cell, but does not add much to the question about complexity and life. It seems that something essential is missing in this regard. Overcoming this situation may require a new theoretical framework which builds on the established facts and knowledge, but puts them into a new context. As a result it can complement the existing evolutionary theory by clarifying why and how matter starts to organize itself into living beings of increasing complexity. An example for what I mean is classical physics and quantum physics. While the latter has offered the theoretical basis to describe matter on the atomic level, it has not replaced the laws of mechanical physics, which are still relevant to describe the behaviour of matter on the macroscopic level.

How is it possible to come to such a more comprehensive understanding of life? Personally I am convinced we are facing a purely conceptual problem. In other words, the unsatisfying understanding of biological complexity is not the consequence of still missing data or some decisive experiments which have not been made yet. It is rather necessary to think about a radically new approach, one which opens a new way to order and interpret the known facts. This is good news for somebody like me, who is not part of the scientific establishment. For a theoretical enterprise of this kind there is no need for a lab and research funds. All you have to do is to find a quiet place and start thinking how to explain the established facts in a radically new way. To be successful one needs enough time and freedom to think out of the box. In this regard it may even be an advantage not to be a part of the scientific establishment without any time pressure to publish or to seek funds. But what exactly can be expected as the result of such a theoretical exercise? In the best case it can detect a law-like process, which explains why and how life can emerge from 'dead' matter and become increasingly complex. And this is exactly what this book is about, the discovery of a law-like process of complexity increase as an essential part of the evolution of life.

At this point I want to address myself directly to you, dear reader. Obviously the following text demands your willingness and openness to look into a radically new approach quite different from the established theory. I am fully aware that this may not be easy, given the fact that today all theoretical thinking in biology is heavily focused on and around the concept of genes. But it is also clear that the development of a radically new approach can only happen by overcoming the established associations and beliefs. Therefore I want to ask you, please set your mind free for a different view on life and evolution. At the end you may then decide if the proposed theoretical framework contributes to a better understanding of the phenomenon life or not.

While starting to write this book, I became aware that the task has two different aspects. The first has to do with the presentation of the content, this in the sense of how exactly a new theory about complexity and life should be presented, so that the reader is able and willing to follow. Here I can only hope that I found a way of not being too theoretical and boring, but rather meaningful and inspiring. The second task has to do with the format of the book, this in the sense of in what detail the existing literature about the subject should be reviewed. When it comes to complexity and life the terrain to cover is huge, it stretches from chaos theory and thermodynamic, to evolution and ecology. To do justice to all these aspects is, I believe, very difficult if not impossible. Therefore I decided to avoid the risk of losing focus, and to concentrate on the essential. This means, I will only refer to publications that are directly relevant for my arguments. As a result some readers may find the text not detailed enough, while for others the contrary may be true. In any case, if I have not referred to the work or idea of somebody else who deserves to be mentioned, I would like to apologize in advance for it.

Introduction

An unsolved puzzle of modern science is about life and its evolution towards increasing complexity (see e.g. Davies P. 1999 or Heylighen F. 1996). Life on earth emerged about 3 to 4 billion years ago, and there are many good reasons to believe that it started with relatively simple organic molecules forming some kind of chemical network (see e.g. Peretó J. 2005 or Shapiro R. 2007). In one way, until now not understood, certain of these molecules developed the ability to multiply and to pass this capacity on to the following generation (de Duve C. 1995). At this point a fascinating process called evolution was kicked off, which resulted in the emergence of structures of increasing complexity: the first molecules capable of reproducing teamed up with liposome-like structures and formed the first simple cells, which evolved over time into more advanced cells with outstanding biochemical power, the prokaryotes. Then bigger and structurally more complex cells, the eukaryotes, arrived on the scene, which organized themselves over time into first primitive organisms. Those continued the process of biological complexity increase by evolving into a generous biosphere with uncountable organisms, ranging from low to high complexity. And finally, not to forget about 200.000 years ago our own species *Home sapiens* appeared and opened a new, at first glance different chapter in the history of complexity increase, stretching from stone axe to aircraft carrier.

The whole process, from simple molecules to *Homo sapiens*, can be seen as a self-organization towards increasing complexity and provides the subject of this book: why and how is it possible that sterile, inorganic matter starts to organize into a manifold biological diversity including ourselves? Or in other words, this book deals with the cause and the mechanism for the self-organization of matter into living beings of increasing complexity.

Biological complexity increase demands an explanation

In the physics classes at school we learned that a system left on its own will inevitably go from a state of higher order to one of lesser order, or as expressed by the famous second law of thermodynamics: the entropy of a closed system will inevitably increase over time (see e.g. Schneider E.D. and Kay J.J. 1995). This law is exceptionally well confirmed and is in line with our daily experience, where things do not self-organize into a state of higher order. However, when it comes to the evolution of life, everything seems to go into the opposite direction. Even if we know that life is not violating thermodynamics, the self-organization towards forms of increasing complexity, and therefore higher order, is an astonishing and exceptional characteristic of life from the physical point of view (e.g.

Davies P. 2003). Very surprisingly however, this subject is not very present in the life sciences. In modern textbooks about biology or biological evolution you may not even find the phenomenon of an emerging complexity mentioned, let alone a discussion of the question of how it fits our physical understanding. In order not to be misunderstood, it is not my intention to accuse biologists as a whole of ignorance in regard to this important aspect of life and evolution, which is clearly not the case. However, for me it is astonishing and sometimes frustrating to see the discrepancy between the relatively low attention given to the phenomenon by the mainstream in bioscience on the one hand and its fundamental importance on the other.

What are the reasons for this discrepancy? In my personal opinion the readiness to put the question of biological complexity not very high on the scientific agenda may have to do with the inevitable discussion about its implication. Here I mean that probably the first association coming to mind is that between complexity and progress. Does the emergence of biological complexity mean that evolution results in some kind of progress and is man the result of this development? By conviction, most scientists deny nowadays any special status of mankind in nature, and their answer with regard to progress is therefore no. However, can the conviction that scientifically all life forms have the same importance also mean that biological complexity is without any relevance? The answer is clearly no. It would be completely absurd to deny that evolution followed a trend from relatively simple structures, such as a bacterial cell, towards much more complex structures, such as an oak tree; or to deny the evidence that the complexity of ecosystems has increased during the course of evolution (see e.g. Heylighen F. 1996 or Kauffman S. 1995a). In the same sense it is simply a fact that humanity is very special and outstanding, compared to all other species produced by evolution. Both the obvious trend towards complexity and the emergence of an intelligent species like us demand an explanation. And as long as this explanation is missing, our understanding of life and evolution cannot be complete. And let's be clear, to accept that something qualitatively has happened during the course of evolution is not identical with the approach of seeing mankind as the only valuable crown of creation, as was the case in earlier times. Interestingly, when it comes to the history of civilization the concept of progress is much less taboo and nobody would hesitate to see the development from the Stone Age to Internet as progress. Be that as it may, I personally see it as a necessity to bring the complexity issue back into the centre of modern biology, where it deserves to be.

The meaning of biological complexity

For a discussion of biological complexity, it is obviously necessary to clarify at first what does 'something is complex' actually mean? In science, as in real life, the term complexity can have different meanings (Bennett C. H. 2003; Mitchell M. 2009). For example, the term complexity is used to name a field of study devoted to the process of self-organization, as well as in information technology as the measure of time and space used by an algorithm to solve a problem. When it comes to life and evolution the situation is even more unclear and no generally accepted definition exists (see e.g. in Gregersen N.H. 2003). It is obvious however, that the complexity of living forms shows certain specific characteristics. As an example, elephants and clouds are both complex structures. One fundamental difference between both is that all elephants share a complex form with the same structure, which is based on a common type of organization. On the other hand the clouds in the sky have a shape which is entirely arbitrary and does not show any organizational principles comparable to living forms (Haken H. 1994). This reflects the important aspect that organisms own an organized complexity in the sense that the components must cooperate and interact with each other in a highly specific manner to give the coherent unity. While this may also be true for dead objects, like a computer, the complexity of life has one very specific particularity. The self-organization of complex living forms, such as a human being, needs a very long time. An uninterrupted process of more than 3 billion years of evolution with uncountable reproduction steps was necessary. This means the self-organization of biological complexity is inseparably connected to deep time[1]. By combining this time dimension with the organizational aspect, it is possible to describe biological complexity in the following way: living forms show an organizational type of complexity, which can only emerge over geological periods of time (compare e.g. Bennet C.H. 2003).

Does this particularity indicate that biological complexity has a specific qualitative meaning? Up to now the answer is disputed and more a matter of personal interpretation and conviction (Davies P. 2003). The difficulties to associate any meaningful qualitative aspect with the complexity of organisms may be a consequence of the mentioned time aspect. We are used to seeing complexity connected with the individual organism. But this may not be enough, it may be necessary to consider the weight of geological time periods, in order to develop a better understanding of the possible meaning of complex life forms. As the reader will see, this is exactly what I will do in the following with the result that the complexity of life can indeed be linked to a very specific qualitative meaning.

---

[1] I borrow the term deep time from authors such as Gee H. (2000) and will use it in the following to express the very long periods of time necessary for evolutionary processes to occur.

What is life?

More than seventy years ago, Erwin Schrödinger, one of the leading physicists of the last century, wrote a famous little book with the title "What is Life?" (Schrödinger E. 1944). Schrödinger was struggling with the fundamental difference between living and dead objects. For him the mechanism of genetic heredity looked like the key to finding the answer he was looking for. Since then, knowledge about the molecular basis of life has dramatically increased, and the question about the mechanism of heredity and the molecular carrier of the genetic information has been solved. In addition, many other details about the molecular and biochemical basis of life were discovered at a breathtaking speed. A good example is the successful decoding of the complete genetic information of an increasing number of organisms, including humans. The result is an overwhelming amount of data, and one should expect that we are now in a much better position to answer Schrödinger's question "what is life?". However, despite the unquestionable successes in molecular biology, we still do not grasp what exactly separates a living organism from a lifeless physical object (Murphy M. P. & O'Neill L.A.J. 1995 or Stewart I. 1998 and 2003).

The problem becomes visible in the multiple efforts to come up with a definition of life. To give an example, the definition used by the American Space Agency NASA for the search for life on other planets is the following: "Life is a self-sustained chemical system capable of undergoing Darwinian evolution." Because the sun can also be seen as a "self-sustained chemical system", the decision as to whether something is alive or dead would be based on the criterion of whether it can undergo Darwinian evolution or not. As a consequence, this type of definition may help to decide if a respective system is alive by observing its behaviour over the course of many generations. However, it will not help to answer Schrödinger's question. To undergo a Darwinian type of evolution is without any doubt a central aspect of life, but to say life means it evolves is nothing more than to turn the problem upside down (Davies P. 1999). In this regard I fully agree with those who believe that this unsatisfying situation is purely the consequence of a conceptual problem and not the result of some still missing observational data (see e.g. Eldredge N. 1995, Gould S. J. 2002 or Rose S. 1997).

The problem of arriving at a more satisfactory definition of life may be closely linked to the question how and why biological complexity emerges. If this is true it may be necessary to understand the meaning of biological complexity first, before being able to answer the question "what is life?". And this is exactly what I will do in this book.

The two ways how biological complexity could emerge

Now, if the exact meaning of biological complexity is unclear, what can be said about its emergence? Here two fundamentally different positions are principally possible: the emergence of increasingly complex organisms during evolution may be either totally accidental or may be the outcome of a law-like process[2] towards an increasing complexity (Davies P. 1998).

Obviously the consequences of both positions are very different. Let's start with the view that the increase of complexity during biological evolution is the result of a law-like process. In this case the emergence of increasingly complex biological forms would be the inevitable and logical result of this process. For such a process to take place, it is obviously necessary that the complexity of life has a particular, qualitative meaning. Otherwise, a law-like directionality towards progressively more complex forms is hardly thinkable. The process needs to build on a qualitative difference between different degrees of complexity. The probably most important consequence of this scenario would be the following. If we could 'rewind the tape on earth', or if the same type of evolution would happen elsewhere in the universe, it should follow the same route towards higher complexity. Thus, evolution would again produce a similar result, provided the overall conditions are the same and the law-like process has enough time. Obviously this does not mean that the result is exactly the same, but it should be similar (for arguments why a 'rewinding' would deliver a similar but not identical result see Conway Morris S. 2003). At the end, a law-like process of complexity increase during evolution would also support the assumption that if life has emerged also on other planets, then we are very probably not the only intelligent beings in the universe.

The alternative and opposite point of view is that of complexity as a purely accidental by-product of evolution. Here complex living forms would emerge merely by chance and not via a law-like process. This interpretation is favoured by evolutionary orthodoxy, which sees all living forms exclusively as the result of a gradual adaptation to the environment, driven by nothing else than selection on fitness. The mechanism on which this belief is based is the following (see e.g. Davies P. 2003 or Gould S. J. 2002): starting with a certain type of structure, natural selection acts at random in the sense that it can in principle produce lesser or more complex variations of this starting point. If there is, however, a limit to the degree of simplicity that a living structure can have, this minimum need for complexity acts as a barrier for evolution towards low complexity. Therefore, if evolution starts at the point of minimum complexity it is obvious that only more

---

[2] I found the expression "law-like complexifying process" in Paul Davies' book "The Fifth Miracle", which expresses very well a non-accidental nature of complexity increase; hence I use throughout this book the term 'law-like process of complexity increase'.

complex organisms are expected to emerge. The longer the time, the higher the chance that one appears that will be further away from the starting point, meaning more complex than all the others. As a consequence, over deep time it is to be expected that the average complexity increases. This would be more in analogy with a simple, unidirectional diffusion process, but not the result of a systematic trend given by a law-like directionality. With this scenario the increasing complexity during evolution would have no further meaning or significance (see e.g. Carroll S.B. 2001). The obvious consequence would be that man is nothing more than an accident.

Evidently I would not have written this book if in my opinion the second scenario were true. With regard to the validity of this scenario, the most critical question is, how can the starting point emerge, the minimum of complexity that life needs? If this corresponds to something like a prokaryotic type of cell, this minimum would already be quite complex and its appearance needs an independent explanation (Davies P. 1999). In addition, the whole scenario is based on the assumption that evolution is exclusively driven by a gradual selection on fitness and nothing else. But this evolutionary orthodoxy raises important critical questions in itself, which I now wish to discuss in the following.

Is Darwinism the complete truth?

For the following discussion it is important to differentiate between evolution as such and Darwinism. That life is the outcome of an evolutionary process is very well supported and can be seen as a fact. The way however, how evolution exactly occurs is still controversial (see e.g. Eldredge N. 1995 or Gould S. J. 2002). In this regard the prevailing evolutionary theory is based on Darwin's insights that all life has one common ancestor and that descendants with inheritable variants are sieved by natural selection to retain adaptations with a higher reproduction success. The decisive principle identified by Darwin behind natural selection is that the fittest leaves more descendants behind, which was often expressed in the famous quote of the "survival of the fittest"[3]. Without any doubt, Darwin's concept is of outstanding importance and describes multiple aspects of the evolution of life. As an example, one of its major strengths is the elucidation how to maximize effectiveness under given circumstances (Kauffman S. 2003). However, this concept has also well-known limitations. It is much better suited to explain the modifications of species or forms once they are already existent, but it has difficulties to explain the emergence of a new type of organism or species. In this

---

[3] It needs to be mentioned that survival of the fittest corresponds to a tautology (see e.g. Eldredge N. 1995), in addition, not the survival as such is relevant for evolution, but the fact of reproducing better than others.

respect Kaufman once quoted a late nineteenth-century sceptic who formulated it like this: "It is like archiving an apple tree by trimming all its branches". Even if this criticism may be overdone, it points to the central dilemma of Darwinism (see again Kaufmann S. 2003). And because the emergence of new species is so closely linked with the question of how and why life becomes more complex, Darwinism as such does not help very much to explain the biological auto-complexification during evolution.

About 50 years ago DNA was identified as the carrier of heredity, and the molecular basis of all living forms became evident. Subsequently these fundamental insights into the molecular structure of life were combined with Darwin's concept about evolution in a theory often called Neo-Darwinism. The main characteristic of this theory is to see genes as the genuine object of natural selection instead of organisms. But does this focus on genes change the limitations mentioned with regard to an understanding of how evolution can produce complexity? The answer is no. To speak in mathematical terms, genes are necessary but not sufficient to explain life (see e.g. Conway Morris S. 2003 or Rose S. 1997). One may even ask the question, does life in all cases need DNA or can other life-forms on other planets possibly have a different makeup? I do not want to go further into a critical analysis of Neo-Darwinism with its focus on genes, because this was already done by other, more competent authors (e.g. Eldredge N. 1995, Goodwin B. 1994, Gould S. J. 2002 or Rose S. 1997).

So, does this mean that Darwin is wrong? Let it me put like this: in certain circumstances he provides indisputably an adequate explanation; in others, however, not. This may mean that Darwinism needs to be complemented by additional principles. Let's take for instance the role of the environment. In conventional evolutionary thinking the link between life and environment is rather unidirectional. The role of the environment is mostly limited to being a source of selection on fitness (see e.g. Kauffman S. 2003). In certain situations, this may be an adequate approach towards reality, but not in others. As an example, evolutionary processes that take very long periods of time may need a consideration of the mutual influence and interdependence between the environment and the multiplication of biological forms. As the reader will see later, precisely this will become central with regard to biological complexity. Another example would be the role of sexual reproduction. Why do more complex life-forms normally need two genders to reproduce, while simpler forms are often asexual (see e.g. Wesson R. 1991)? In this regard Darwinism can explain how sexually reproducing species are influenced by natural selection, but the theory is struggling to give a satisfactory explanation of why exactly sexual reproduction evolved in the first place (Eldredge N. 1995). In this book I will show that the split of species into two genders can be closely linked with the need for sustainability during the course of evolution.

Again this means, Darwinism explains how sexually reproducing species become modified after they have emerged. But to understand why life requires this type of reproduction, additional and complementary principles would be needed.

In summary, while Darwin's basic concept is doubtless correct, it looks as if something fundamental is missing. The focus on natural selection and fitness may explain certain aspects of evolution, but it is very probably much too simplistic to explain others, such as the emergence of biological complexity. In other words, it looks as if Kaufmann is right with his remark that Darwinism is the truth but not the complete truth (Kauffman S. 2003).

Towards a new understanding of life and evolution

This book presents a new theory, the *Interaction Theory*, which is intended to contribute to a better understanding of life and evolution, by providing in defined terms and principles an adequate explanation why and how biological complexity can emerge. By doing so, the theory provides also a new answer to the question "what is life?".

Decisive for the new theory is the introduction of a sustainability criterion demanding a balance between living forms and their environment. This approach builds on the insight that unsustainable growth as a result of an exclusive selection on fitness represents a major challenge for life. Due to this sustainability consideration *Interaction Theory* shows that the emergence of biological complexity follows a law-like process, which causes the stepwise self-organization of matter towards increasingly complex biological forms. In this sense the law-like process of complexity increase, as defined in this book, corresponds to an evolutionary algorithm, which transfers an initial state of low complexity into one of high biological complexity. At the end *Interaction Theory* allows the conclusion that the evolution of increasingly complex organisms becomes inevitable wherever and whenever life is emerging. And with brain complexity as an integral part of this development, the emergence of an intelligent species, like ours, becomes an expected outcome, instead of an accidental, almost improbable event.

In the following I will start by discussing and ordering the known facts about life and evolution in a new way, so that relevant insights about the meaning of biological complexity will become visible. Here, it will also become clear why certain types of interactions are so central that they gave the theory its name. Based on the insights gained, I will then derive the mentioned law-like process of complexity increase, which is at the centre of *Interaction Theory*.

Before starting, I would like to end this introduction by mentioning the importance of the subject also outside the biological domain. A radically new understanding of how biological complexity emerges can have implications on the way how we see and understand the role of our own species in nature. This means that it does not only concern philosophical questions such as "are we alone in the universe?", but that the potential implications can go further. So we might come to conclusions about the ongoing auto-complexification of our human society, which might stand in sharp contrast to established convictions. A new understanding of the evolution of life as a result of a law-like process towards increasing complexity may therefore be of importance well beyond biology. At the end of the book I will come back to this point.

Part 1: The formulation of *Interaction Theory*

1.1. The *Interaction Theory*

Interaction Theory as presented in this book is a radically new way that describes biological evolution to increasingly complex forms of life. The new theory is intended to clarify the actual meaning of biological complexity and to provide new answers to the so far unsolved questions of why and how organisms became more complex over geological periods of time. In addition, it also puts essential aspects of evolution and life into a new light, such as the reason why sexual reproduction is necessary and what is the meaning of species. On the following pages I will therefore develop *Interaction Theory* step by step and try to convince the reader of its correctness.

How to approach biological complexity?

What is the best way to come to meaningful insights about why and how biological complexity has emerged during evolution? Obviously the complexity of life in all its details is far too vast to be addressed directly. To come to meaningful insights it is therefore necessary to introduce some kind of model in the form of what can be called an adequate abstraction of reality (Fontana W. and Buss L.W. 1994). This abstraction must reflect the key characteristics of life and evolution, while portraying the essential aspects of reality in a few, well defined terms and principles. In a next step these terms can then be put into relation with each other, so that meaningful conclusions about the emergence of complexity can be derived.

The main criterion to decide whether one specific model is suited or not depends on its results in the sense of: does it explain what has happened during the course of evolution? In the case of *Interaction Theory* this would mean: does it for example help to understand why the phylogenetic tree of life stretches from structurally relatively simple prokaryotes to highly complex forms, such as oak trees and *Homo sapiens*? Or can it also elucidate why the biological diversity in the phylogenetic tree is clustered and not gradually organized? Over and above that, however, an adequate model must make it possible to derive meaningful deductions about the real world, which can then be tested and verified (see Lloyd E.A. 1994). An example would be a testable prediction about the emergence of new species.

But what can an adequate abstraction of reality look like? In this respect it makes sense to consider that the emergence of biological complexity is closely linked with the phenomenon of life as such. To come to an adequate model with regard to biological complexity it will therefore be crucial to identify an adequate description of life that makes it evidently different from 'dead' objects. From here on, the desired model can then be built. The whole endeavour to come to meaningful insights about the why and how of biological complexity may therefore depend on this choice.

Now, what makes life so fundamentally different? From the viewpoint of evolutionary orthodoxy the capability for reproduction and heredity based on genes are normally seen as the decisive key features of living forms (see Dawkins R. 1976 or Williams G.C. 1966). This focus on reproduction and genes does not, however, cover the important particularity of biological complexity, namely that it has emerged over deep time. Therefore I want to choose a different and radically new approach. I will put the emergence of generation sequences continuing over deep time into the centre.

The generation sequences of life

In general, life and evolution are linked to the fate of individuals and their genes by focusing on the individual capability to reproduce. However, this does not acknowledge that biological complexity does not emerge in a single generation, but is the result of a slow process over very long periods of time and uncountable generations following each other. For the discussion of complexity increase during the evolution of life it seems therefore more adequate to consider this long process, instead of focusing on individuals. How can this be done? The solution offered in the following is to connect the discussion about biological complexity with generation sequences. Here I mean the following: life has the characteristic of manifesting itself in the form of uninterrupted sequences of generations proceeding over long, geological periods of time. All individual forms living today, be it the uncountable bacteria, alga and protozoa, as well as plants, mushrooms, animals and humans, all have in common the fact of being the outcome of a very long sequence of generations stretching back into deep time to the origin of life. And because it can be assumed that all life on earth originated from a common ancestor (Doolittle W.F. 2000), all generation sequences would have started from the same point. Based on this insight, life and the emergence of complexity can be seen as a process and described in the following way:

- Life means the emergence of generation sequences, which start from a common point, proceed over deep time and result in living forms of different complexity.

These generation sequences can be seen as a kind of chain reaction, driven by a reproduction event in each and every generation. On earth, the chain reaction was most likely kicked off more than 3 billion years ago, and continues to this day (for an overview about evolution, see e.g. De Duve C. 1995). And no foreseeable end exists, because at least some of today's generation sequences may continue as long as conditions exist that allow the reproduction of the respective forms. But not only this, the generation sequences have one other, very surprising characteristic. During their course through time the living forms carrying the generation sequences of life have changed and became more complex. At the beginning, while starting with the last common ancestor of all living forms, all generation sequences were on the same, very low level of complexity. However, this uniformity changed dramatically. In the case of prokaryotes the complexity had relatively rapidly reached a point after which no dramatic changes happened any more over billions of years. On the other extreme, the sequences leading for example to humans have experienced over the course of evolution a progressive complexity increase. Other living forms can be seen somewhere in-between the two extremes. In the following I will try to understand why and how generation sequences can proceed over deep time and cause this pattern of biological complexity.

Interdependence between living forms and environment

The focus on generation sequences has the effect that it becomes crucial to see life and evolution under the aspect of an ongoing process, which needs to be sustainable over deep time. This results in fundamental differences with current evolutionary mainstream thinking. Mainly the interdependence between living forms and their environment does acquire a different meaning and dynamic. Before discussing these differences in detail, I want first to point out some aspects regarding the role of the environment in traditional evolutionary thinking.

Thanks to Darwin, the basic principles of evolution seem to be simple and clear. On the other hand however, their interpretation is often influenced by the relevant zeitgeist (see e.g. Fox Keller E. 1995). For example, in Darwin's epoch, when a bad harvest caused a general famine, nature represented a threat, which demanded the constant struggle to survive. Not surprisingly this is reflected in Darwin's theory, which presents evolution as a history written with teeth, blood and claws (Darwin C. 1859). And obviously, in such a world only the fittest can survive. In a similar way the role of natural selection can be seen as influenced by the zeitgeist of the Victorian epoch. Strict and authoritarian, natural selection rules in absolutism and decides on the fate of all living beings struggling with a merciless environment. Unfortunately this simplistic view of evolution has not completely disappeared since Darwin. Today, in the era of a global economy and its obsession with profitability and performance, the fixation on fitness seems even to be stronger.

An understanding of evolution in the way mentioned ignores completely the fact that living forms do also represent a potential threat to a fragile environment, which may become irreversibly exhausted by fast and self-enhancing growth. Here global humanity gives a concrete example of being directly confronted with the ecological consequences of its fast and strong growth over the last decades. To come to a more adequate interpretation of evolution it would therefore be important to consider much more the close interdependence between living forms and their environment. And this does mean organisms need also to be seen as potential multiplication bombs. They react with exponential growth, whenever favourable conditions appear. One tiny little bacterial cell, much too small to be visible with the naked eye, is able to divide under optimal conditions each and every 20 minutes (Cooper S. and Helmstetter C.E. 1968). As a result the total earth surface can theoretically be covered with a thick layer of bacteria after only 36 hours of exponential growth (Pianka E. R. 2000). Obviously, these dramatic consequences of self-accelerating growth, which can destroy the environment and exhaust resources, must be taken into consideration. This leads to the following conclusion: to understand why life became more complex during evolution demands a holistic view of living forms as potential victims of environmental selection as well as potential perpetrators, which can irreversibly damage the very same environment. Or less prosaically, it needs to be considered that not only the environment influences and alters living forms, but that also the environment becomes influenced and altered by the ongoing multiplication of living forms. This is exactly what *Interaction Theory* does, by closely linking evolutionary processes with sustainability.

Regarding the influence of the environment it is also necessary to take into consideration that the environmental conditions are not always changing in a gradual or linear manner. Thus, from the generation sequences point of view it is necessary to anticipate that over deep time the environment cannot be stable and does sometimes change abruptly and dramatically (see e.g. Lenton T. & Watson A. 2011). When such sudden and dramatic environmental changes occur, the reactive adaptation by the affected living forms may not be possible and mass extinction occurs. A well-known example of this kind of event happened very probably about 65 million years ago, when a meteorite wiped out the ecosystems dominated by dinosaurs and opened the way for mammals and, in the end, *Homo sapiens* (see again e.g. Lenton T. & Watson A. 2011). Later I will show that the complexity increase in evolution is driven by this type of catastrophic event and change.

Sustainable generation sequences

The focus on generation sequences and the insight that living forms may exhaust and irreversibly change the environment with self-accelerating growth serves as a key insight for the following discussion of complexity increase. To approach evolution under this aspect means that the long and ongoing generation sequences of life occur in an environment that can become exhausted or destroyed by growth. As the reader will see, it turns out that here lies the key for a better understanding of biological complexity. This brings me to the starting point of *Interaction Theory*, the introduction of a sustainability criterion demanding a balance between living forms and their environment. This demand results in the following key question:

- As life means the emergence of generation sequences which proceed over deep time, why and how can a biological auto-complexification occur, if living forms represent potential multiplication bombs in an environment with exhaustible resources?

To start with this question has the advantage that the evolution of complex life becomes tangible. And it will turn out that the answer on how a sustainable multiplication of living forms is possible in an irreversibly exhaustible environment will provide the solution as to how and why biological complexity increases over time, namely by a law-like process.

To direct the reader to this conclusion, I will proceed as follows: first I will identify the key insights essential for generation sequences to travel through deep time in a world with exhaustible resources. Based on these insights, I will then answer the question of why and how in some generation sequences the respective forms become more complex over deep time. This will then lead to a law-like process of complexity increase at the end of the first part of this book. In the second part, I will then discuss the course of evolution on earth, from its beginning up to the appearance of *Homo sapiens*, under the aspect whether it has indeed followed the identified law-like process of complexity increase. This means I will use the gained insights and conclusions in order to show that they are in line with what has happened in reality. At the end the reader may decide whether or not *Interaction Theory* is appropriate to explain the emergence of biological complexity during the evolution of life.

1.2. The framework to study generation sequences

In the following I will define the conditions which are necessary for generation sequences to travel through deep time. This provides the framework or model for the subsequent discussion of why life results in living beings of different complexity.

Multiplying forms, generation sequences and complexity

In biology life is normally closely linked to the appearance and characteristics of cells. On the other hand, from the point of view that the self-organization to higher complexity is an ongoing process over deep time, the cell is merely more than an intermediate step. Evolution started very probably with relatively simple molecules evolving into more complex molecular entities, which then aggregated into cells, of which over time some organized into multi-cellular organisms and resulted in the manifold diversity of herbal and animal life (see e.g. De Duve C. 1995). With the arrival of *Homo sapiens,* this process of complexity increase finally entered into a new phase, which is characterized by a cultural, scientific and economic evolution towards more and more complex societies (see e.g. Tattersall I. 1998). To understand why life on earth became increasingly complex, I will therefore focus on the process as such and not on particular biological entities, such as cells. As a consequence, the expression form or forms will be used throughout this book as synonym for all possible entities of life such as cells, organisms or the molecules at the beginning of evolution. Generation sequences can therefore be carried on by different types of forms, ranging from molecules to human beings. This simplification makes it easier to concentrate on the process and to ignore, at least for the moment, the differences between the particular forms.

With regard to respective forms it is further assumed that they are distinct entities with the capability to reproduce, either alone or with a partner. Further, they do not need to consist of one physical unit; they can also consist of different parts forming some kind of organizational structure, such as a chemical network or an ant colony. It is essential that the subparts of such a composed form can no longer reproduce independently, but need the whole entity. This means generation sequences causing forms can show what I called in the introduction the organizational aspect of biological complexity.

What else can be said about generation sequences causing forms? Most important, they must follow Darwin's concept that descendants with inheritable variants are selected, to retain adaptations with a higher reproduction success (Darwin C.

1859). In other words, generation sequences are carried on by forms with the capacity of reproducing in a way that the next generation shows inheritable variations and is therefore not 100% identical with the previous. The variations or differences between the individual forms in each and every generation then influence the fact that some of them can carry their sequence further on, while others not.

Another important term to clarify is complexity itself. In the introduction I have mentioned that the complexity of living forms shows the particularity of an organizational type of complexity, which is emerging over deep time. These aspects are, however, not very practical with regard to a comparison of the complexity of individual forms. Therefore I will use the following definition as a starting point: the complexity of a form means its structural complexity, in the sense of the variety and quantity of its structural details. In other words, the more complex a form is, the more structural details it owns – for example, eukaryotic cells have more structural details than prokaryotes, or a multi-cell organism more than a single cell etc.

I have to admit that this is a rather rough definition. It offers, however, the important advantage that the complexity of forms becomes tangible. So it allows the conclusion that the structural complexity of a form must have a direct influence on the reproduction rate. Hence, simple forms can reproduce faster than complex ones, because the number of structural details that need to be replicated during the reproduction process is lower. As a consequence, the time between two following generations must become longer with increasing complexity. This in itself is remarkable, because living forms should have a selective advantage if they reproduce fast. Nevertheless, a positive correlation between complexity and reproduction rate corresponds largely to what can be found in nature, with prokaryotes reproducing faster than eukaryotes, single-cell organisms faster than multi-cellular organisms etc. (Bonner J.T. 2006). In this regard it is necessary to mention that biological forms can have different types of reproduction, which influences the possible number of progenies per generation. In addition big organisms, such as elephants, need per se more time to grow, and therefore to reproduce, than much smaller forms, such as mice. For the moment however, I want to ignore these aspects, but will come back to them.

In conclusion, the link of complexity with the number of structural details serves as a first, preliminary definition. In the following, it will become clear that the complexity of living forms can be linked with another, very distinctive qualitative meaning.

Generation sequences need multiplication

Forms able to cause generation sequences need the capability to reproduce. From now on however, I will exclusively use the term multiplication and not reproduction. The reason is that reproduction implies making a copy of something, while multiplication implies the aspect of making more. The term multiplication expresses therefore much better the fact that living forms are potential multiplication bombs. In addition, in a cell the only structure to be copied is DNA. All other parts of the cell need some kind of multiplication process before or during the cell division. Good examples for this are biological membranes as crucial parts of cells (e.g. Watson H. 2015).

Multiplication means that under favourable conditions more of a respective form is produced than destroyed. Thus an individual form becoming multiplied will increase its number. In chemical terms the multiplication process can be described as a disequilibrium state with one form (F) plus x-times building blocks (BB) resulting in at least two or more (n) forms:

$$F + x\ BB \rightarrow n\ F$$

From the physical-chemical point of view this is crucial, because individual forms, which are locked in a thermodynamic equilibrium state between formation and destruction, cannot cause ongoing generation sequences continuing over time (for thermodynamic aspects of evolution see Schneider E.D. and Kay J.J. 1995 or Stewart I. 2003). Thus, for evolutionary processes to happen, the underlying generation sequences need to be driven by multiplication or making more events (regarding 'making more' and reproduction see e.g. Eldredge N. 1995). Otherwise biological forms cannot carry generation sequences through deep time. In the following I will therefore use the expression multiplying form for the carriers of generation sequences.

The environmental conditions for generation sequences

To proceed over deep time, generation sequences need suitable environmental conditions. For this reason, I will now introduce an adequate abstraction of the environment, in order to make the otherwise much too complicated situation in the natural environment tangible.

The first simplification with regard to the environment concerns the two factors entropy and energy. Evolutionary processes demand an open energetic system due to thermodynamic reasons (Davies P. 1998 or Schneider E.D. and Kay J.J. 1995).

In the following it is therefore assumed that the necessary energy and entropy to sustain the course of generation sequences over time is available from sources such as geothermal heat or solar energy. An example would be the production of activated molecules by sunlight, which can then directly or indirectly deliver the energy needed by the multiplying forms. Those forms which are able to use this type of energy sources directly would correspond to autotrophic organisms, the others to heterotrophic (Campbell N.A. et al. 2008).

Beside energy, the multiplication of a form needs other resources too. Examples would be substances serving as building blocks or base material, such as essential amino acids, fatty acids or certain sugars. This would also comprise resources such as water or space to live and grow. Common to all kinds of resources is that they must be shared with other forms and that their availability impacts the multiplication success (see e.g. Stearns S.C. and Hoekstra R.F. 2005).

In reality life is, and always was, confronted with exhaustible resources (see e.g. Pianka E. R. 2000). Therefore, I will introduce as an important characteristic of resources that they are exhaustible. This in turn has as consequence that generation sequences have to deal with the fact that essential resources might become exhausted or destroyed by overconsumption. Here, two alternatives have to be considered. In one case, a resource can become irreversibly exhausted or destroyed. An example would be the prey of a predator - if too many foxes have hunted down the last hare, this 'resource' is gone forever. In the second case, the resources can regenerate after exhaustion or destruction by overconsumption. Examples are soil minerals essential for plant growth or drinking water for animals. For complexity increase during evolution this means: generation sequences are confronted with exhaustible resources which have either the capability to regenerate after their depletion by overconsumption, or to go irreversibly lost. As a consequence of this resource situation, the environment cannot tolerate an unlimited number of forms, or respectively, the multiplication cannot result in an ongoing, unlimited population growth.

One further, necessary assumption regarding the environment is the following. The environmental conditions necessary for generation sequences to continue over time correspond to a combination of different, distinctive factors. And obviously, the exact conditions necessary for the multiplication can differ between different kinds of forms. The description of how to cultivate a specific strain of micro-organisms, with specifications such as the composition of the medium, the temperature and pH of the solution, would be an example of what is meant. The factors relevant for the multiplication of a form can be further divided into physical-chemical, such as the temperature or essential substances in the medium, or they can be biological, such as predators, parasites, symbiosis partners or other forms in competition for space or food. For example, a certain prey would correspond to a biological factor that the environmental conditions must provide, in order to be suited for the multiplication of a particular predator. Thus, the factors influencing the multiplication of forms can either be abiotic or biotic.

At the end of this section I want to mention the term 'niche' and its link with generation sequences. A niche stands in the following for the specific environmental conditions that allow respective generation sequences to travel through time (for traditional niche definitions see e.g. Campbell N.A. et al. 2008). As a consequence of what was said before about the environment, a niche can therefore be described as a set of abiotic and biotic factors to which a particular form is adapted, so that its generation sequences can continue over time. In the following, the environment will therefore be portrayed as a mosaic of environmental conditions, which can either serve as niche and allow a particular type of generation sequence to continue over time, or not. Later I will come back to this point and discuss niches in more detail.

1.3. The fate of generation sequences

Hitherto I have looked at several aspects and conditions relevant for generation sequences. In the following I will consider the competition between the different variants of a multiplying form and what it means for the course of generation sequences. In this regard different terms become important, such as multiplication success, selection on fitness and a very central quality of generation sequences, which I call the probability of multiplication.

The consequence of an ongoing multiplication

Generation sequences are the result of multiplication events which produce descendants with heritable variants. The different variants of a multiplying form are the subject of what Darwin called natural selection (Darwin C. 1859). The decisive quality in this regard is the individual multiplication success of each variant under the given environmental conditions. The multiplication success corresponds to what is generally called the reproductive success. It can be defined as:

$$Msuc = N / t \ (EC)$$

N corresponds to the number of descendants per time t under the actually given environmental conditions EC. Regarding generation sequences, this means that in each generation the variants are selected on their individual multiplication success. As a result, those variants become dominant over the generations, which produce relatively more descendants per time. If the environmental conditions are favourable for the multiplication of the respective forms this will cause self-accelerating growth. In other words, this type of selection provokes a general growth contest, for which bacteria are the typical example (for reproductive success and growth rate see e.g. Stearns S.C. and Hoekstra R.F. 2005).

The fact that the selection on the multiplication success goes along with a general growth contest must be of central importance regarding the question how generation sequences can continue over time. By this I mean, generation sequences result from multiplication in the sense of a making more event, while on the other hand they depend on exhaustible resources. This has the inevitable consequence that generation sequences risk exhausting their resources through growth. And this in turn has potentially negative implications for the environmental conditions on which the generation sequences depend. Some pages before I have used in this regard the expression multiplication bomb. The following impressive example from Ian Steward (1998) illustrates the devastating impact of such a bomb: to replicate something means that the number of copies increases linearly, e.g. if a photocopier copies 1000 documents per day, this will produce about 360.000 copies in a year. If however, the machine were able to multiply, in the sense of making more of itself, the number increases exponentially. This means that if the photocopier makes only one copy per day of itself, then after one year, there would be the incredibly high number of 2^{365} photocopiers or roughly 10^{110}.

Consequentially, any theory about generation sequences must consider the potentially devastating effect of an ongoing multiplication in a world with exhaustible resources. It is obvious that only those generation sequences can continue over time that do not irreversibly exhaust or destroy their environment. This allows me to rephrase the previously introduced, central question:

- If certain environmental conditions are favourable for the multiplication of a specific form, how can the corresponding generation sequences continue over deep time, without irreversibly exhausting or destroying these environmental conditions by growth?

Obviously this implies that generation sequences need to be in a balance with their environment. Hence, in order to understand evolution from the generation sequence point of view it is necessary to consider sustainability aspects and that is exactly what *Interaction Theory* does.

As mentioned previously, to view the evolution of life primarily under sustainability aspects stands in contrast to the belief that evolution can be understood by fitness considerations only. For example in certain models, the environment is

24

reduced to so-called fitness landscapes and evolution reduced to the question: how living forms can climb on the fictitious fitness peaks (see e.g. Kaufmann S. 1995a). In comparison herewith, the focus on the above question reflects the mentioned radical conceptual shift, which puts the mutual interdependence between living forms and environment into the centre. It will soon become clear that this will lead to a new understanding of biological complexity.

Generation sequences and selection on fitness

The environment selects forms on their multiplication success. The individual multiplication success of different variants of a particular form can vary under the given environmental conditions (according to the equation Msuc (EC) = N / t). Because the environment can only sustain a limited number of generation sequences, the variants with the relative greatest multiplication success would replace the other variants over time (see e.g. Stearns S.C. and Hoekstra R.F. 2005). As a consequence, the multiplication of multiplying forms becomes over time adapted to the given environment. From the generation sequences point of view adaptation means two things. Firstly, a variant may have a higher survival rate under the respective environmental conditions. An example would be an increased temperature resistance in an environment marked by temperature challenges. As a result the generation sequences originating from this variant have a higher chance to continue over time and not to get stopped. Secondly, the adaptation can result in a faster multiplication relative to others. This means the generation sequences of the corresponding form show relatively more branching events per time. In the following I will describe this fact as a rising multiplication potential. In both cases, the number of the corresponding generation sequences can increase relative to others and eventually prevail. In conclusion, multiplying forms experience a permanent selection on their multiplication success. It favours the generation sequences of those variants, which are best adapted to the environment and produce therefore relatively more offspring per time. For this type of selection I use the term selection on fitness.

In biology, the term fitness stands normally for the ability of an individual both to survive and to reproduce (Stearns S.C. and Hoekstra R.F. 2005). For a sexually reproducing species this means that the relative fitness of an individual describes not only its ability to survive, but also to gain a reproduction partner for common offspring. However, in the following it will become clear that from the generation sequence point of view both aspects need to be separated. Therefore I will use the term fitness only for the possible number of offspring that a multiplication forms can have, due to its specific environmental adaptation and its multiplication potential, as defined above. This means that the fitness of a sexually reproducing species does not consider those aspects which have to do with the need to gain a reproduction partner. For this I will introduce in the following a second type of selection, which needs to be differentiated from selection on fitness.

The so-called Spiegelman experiment gives a concrete example for the consequences of selection on fitness on the branching frequency of generation sequences. With a relatively simple experimental setting Spiegelman (1970) illustrated the effect of an exclusive selection on fitness on multiplying forms, in this case RNA molecules. The trial confirmed that if multiplying forms are selected over the generations exclusively on the number of offspring they can produce per time, then those variants which multiply the fastest under the given conditions will finally prevail. In the experiment RNA molecules of one type and certain length were constantly replicated. The respective generation sequences started in one test tube and after the replication reaction had happened for some time and the density of RNA molecules had increased a small fraction of these molecules were transferred into a fresh tube, where the replication reaction continued. This procedure was repeated multiple times. Each transfer into a new test tube allowed the restart of the multiplication in a freshly regenerated environment. After many transfers the trial ended not surprisingly with RNA molecules shortened to a minimum in length. They just conserved the sequences absolutely necessary for their replication and achieved herewith the fastest multiplication under the given circumstances (see e.g. Maynard Smith J. and Szathmáry E. 1995).

In the light of what was said above it is obvious that the striving for fitness can only be part of the story of how generation sequences proceed over deep time. Under environmental conditions favourable for the multiplication, selection on fitness will result in a self-accelerated or exponential growth, which must cause an exhaustion of resources and the consequence that the multiplication will stop. This dynamic is well known from micro-organisms. They show an oscillating, rollercoaster-like growth pattern, marked by the ongoing change between periods with exponential growth and stagnation (e.g. Cooper S. 2012). In the case that the multiplying forms are molecules, it would mean that the multiplication reaction would end in the deadlock of a thermodynamic equilibrium situation, which is reached when the number of newly formed molecules per time is equal to the number of degraded molecules. In the case of complex organisms, the exhaustion of essential resources may be irreversible and cause the extinction of the respective species. This allows the conclusion that irrespective of the type of multiplying form, an exclusive selection on fitness will inevitably cause self-induced changes in the environment, which bring the multiplication of respective forms to a stop. As said, these negative consequences of growth must be considered as a critical barrier for generation sequences to continue. So it would be necessary for molecular forms, e.g. at the beginning of evolution, not to get trapped in a thermodynamic equilibrium (for thermodynamic and complexity see again Schneider E.D. and Kay J.J. 1995 or Stewart I. 2003). In the case of organisms it would be necessary to avoid the irreversible exhaustion of essential resources. In this regard it is striking that higher organisms show in general a much less fluctuating population dynamic compared to the rollercoaster-like growth pattern of micro-organisms (see e.g. Campbell N.A. et al. 2008).

In summary: for generation sequences to travel over deep time, the following potential conflict is important. On the one hand, the multiplication that drives the sequences must be in a sustainable balance with the environment. On the other hand an ongoing, exclusive selection on fitness results in an exhaustion of resources.

The probability of multiplication

In the following I will discuss how generation sequences can last in a sustainable way without irreversibly exhausting or destroying the environment. For this discussion it is necessary to capture the mutual influence between environment and multiplication. Terms like multiplication success or fitness are for this purpose not very helpful, because they relate to the individual form. For the mutual dependence between multiplication and environment it is rather necessary to consider longer periods of time with many generations. Therefore I will introduce a new term which reflects this interdependency much better: the probability of multiplication. This term will become very central for the understanding of why and how complexity increase happens during evolution.

The probability of multiplication refers to generation sequences and not to the individual form. It stands for the average probability by which an individual generation sequence of a particular form, such as of the bacteria *Escherichia coli* or *Homo sapiens*, can continue for long periods of time without getting stopped. A long period means much longer than the average generation time. As an example, each individual bacterium or human is the outcome of a generation sequence, which goes back to the beginning of evolution. But by far not all bacteria or humans that lived, for instance, 1,000 years ago, belong to one of the generation sequences that were not stopped during this period of time. Therefore, the probability of multiplication (PM) of a certain form expresses the ratio between the number of existing generation sequences and all generation sequences which are theoretically possible, if no sequence had been stopped in the corresponding period. This can be expressed as follows:

$$PM = GS\ (t)\ /\ GS\ (total)$$

GS (t) stands for the number of generation sequences actually proceeding over the longer time period t, and GS (total) for the number of all generation sequences which could have theoretically been possible in this time period.

To make this more concrete, let's take the comparison of the probabilities of multiplication of the generation sequences of *Escherichia coli* with those of *Homo sapiens*. The numbers of individual forms that live at a particular time in a particular place correspond to the number of generation sequences existing at this respective time and place. All these generation sequences can in principle continue and become more by branching. For bacteria this means that one generation sequence branches into two with each cell division and for human couples, with an average of, let's say, three children, two sequences become three. Together with the average generation time this allows to determine the average branching frequency per time for the particular type of generation sequence. With this it is possible to estimate the number of generation sequences that the corresponding population would have theoretically been able to produce during a longer period, such as 1,000 years. This number can then be compared with the number of generation sequences which did not become stopped prematurely, but actually stretch over this period of time[4]. The respective relation delivers the probability of multiplication. In the example the number of actually "surviving" generation sequences divided by the total number of all theoretically possible sequences may be x for bacteria, and y for humans. The numbers x and y correspond to the respective probability of multiplication of each form and can be compared. It may turn out that y is much higher than x. This means, the probability of multiplication for humans is y/x times higher compared to *E.coli* over the chosen period of time. As a consequence it can be said that the average individual generation sequence of humans has a much higher likelihood to travel through time than the average generation sequence of *E. coli*.

Later I will show that the value of the probability of multiplication correlates with the complexity of forms. This means that if the probability of multiplication of all living forms is compared, prokaryotes are at the lower end and humans at the high end. The highest theoretically possible probability of multiplication is one. In this case, all generation sequences can continue, without the risk of getting stopped. Accordingly, the probability of multiplication becomes zero if it is foreseeable that the corresponding multiplication will no longer be possible.

---

[4] This number would be roughly the same as at the beginning of the comparison, if the environment did not significantly change and the population stayed in balance with its environment over the selected time period.

Now, what does the probability of multiplication actually mean? It simply reflects the average chance per individual generation sequence to become hit by events that will stop it. In other words, the probability of multiplication of a specific multiplying form is dictated by all possible causes stopping generation sequences from continuing. The origin of these causes can be external or self-made. External stands for both negative events hitting the individual form, such as a fatal lightning strike or a deadly encounter with a beast of prey, as well as negative environment changes impacting all respective forms simultaneously, such as a dramatic climate change. In contrast, self-made describes the negative impact of the multiplication on the environment, via an exhaustion of resources or a destruction of the environment as a consequence of the multiplication. What is important is that irrespective of the different kinds of negative events, at least some generation sequences continue. Otherwise the multiplication of the respective form is not sustainable and the probability of multiplication becomes zero. Hence, generation sequences require a positive probability of multiplication, so that they can proceed over time. Its exact value reflects the risk by which the progression of an individual generation sequence becomes stopped by all sorts of external and self-made causes. In other words, the higher the probability of multiplication, the higher the likelihood that the individual generation sequence of the corresponding form can continue over time.

In summary, the introduction of the probability of multiplication marks a key difference with evolutionary orthodoxy. The term considers the close interrelationship between the multiplication of living forms and their environment. For a positive probability of multiplication the generation sequences need to be in a sustainable balance with the environmental conditions to which they are adapted. This balance is influenced by external and self-made factors. Based on these insights I will now try to define how environmental conditions must look like in order to serve as niches for generation sequences.

1.4. The niche requirements of generation sequences

Next I will discuss the term niche. As said, compared to the general understanding of niches, the interrelationship between the environment and generation sequences is seen as central. Thus, from the *Interaction Theory* point of view environmental conditions can only qualify as niches, if they are able to sustain generation sequences.

Now, what can be said about environmental conditions suited for generation sequences? Here it should be relevant that selection on fitness favours forms with more progeny per time. Those will sooner or later supersede slower multiplying forms. Therefore it can be assumed that a particular niche is occupied by one type of generation sequence. This would be those which are caused by the multiplying form with the relatively highest multiplication success under these conditions (see e.g. Maynard Smith J. and Szathmáry E. 1995). As said before, it is also critical that a niche can sustain generation sequences. To understand what both aspects mean, I will first discuss the already mentioned external causes which can stop generation sequences.

A basic request for a niche would be that the generation sequences of forms adapted to the respective environmental conditions should not be too frequently hit by external causes, so that all sequences eventually end. As previously mentioned, external causes can be divided into two kinds. Firstly, negative events which stop individual generation sequences, such as a fatal encounter with a predator. Secondly, general environmental events or changes which hit all corresponding forms simultaneously. An example would be a deterioration of the climate. The prospect for the individual generation sequence to either carry on or to get stopped depends obviously on the frequency and magnitude of these negative events or changes. If they are too frequent or strong all generation sequences can be wiped out. In addition this includes that the respective environmental conditions as such do not simply disappear after some time, because they are unstable or in any other way not suited to exist over longer time periods. From the generation sequence point of view, niche would have to fulfil these basic demands. In summary this means that niches must provide environmental conditions which are sufficiently reliable and steady, so that the respective generation sequences can continue over time. And sufficiently means that the frequency or strength of external events or changes does not exceed a critical level, because above this level all respective generation sequences will end.

With regard to such a limitation time must play a decisive role. The reason is that irrespective of a particular niche situation, the likelihood for negative events or changes increases over time (see e.g. Gribbin J. 2004). It makes therefore a big difference, if the demand for reliability or steadiness refers to a period of 10 minutes or 10 years. And this in turn means that the generation time of a form becomes important. For example, the generation time for humans is today about 25 years and for bacteria such as *E.coli* close to 25 minutes under optimal conditions. As a result, the number of branching events per generation sequence and time varies dramatically between both forms. Consequently these differences become important with regard to the demand for reliability and steadiness. Next, I will therefore discuss niches under this aspect.

Environmental conditions that can serve as a niche

From the generation sequences point of view, niches need to be reliable and steady. This does not exclude that in the course of a generation, a part of the generation sequences can become stopped. The frequency or strengths with which this happens needs however to stay under a tolerable level. Consequently, environmental conditions cannot serve as a niche for a multiplying form, if in the course of a generation the frequency or strength of negative changes and events is so high that all generation sequences stop. This can be expressed as follows:

- A niche corresponds to an environmental situation that allows multiplying forms to adapt and continue their generation sequences, because it provides sufficiently stable and reliable conditions, so that not all corresponding generation sequences are stopped during a generation time.

This understanding of a niche differs from the conventional view. Instead of individual forms and their specific environmental competence at a particular point in time, niches are seen under the aspect that generation sequences depend on their relative stability and reliability.

The reference to the generation time considers that the stability and reliability of environmental conditions depend on the respective length of time of a generation. This has the consequence that environmental situations need to be relatively more steady and reliable, if the generation time increases. The reason is that fast multiplying forms can in principle tolerate more negative, generation sequences stopping events or changes per time than forms with a longer generation time. Imagine for example two forms, one with a generation time of 1 year and the other with 10 years. For simplicity let's suppose that both forms multiply in each generation by doubling. A sustainable multiplication demands in both cases that the number of generation sequences stays in balance with their environment. However, the first type of generation sequence can branch much more often. Within 10 years, a single generation sequence can theoretically split into more than 1000 generation sequences, while the number of the second kind of generation sequence has merely doubled. Consequently, the more frequently branching generation sequences can tolerate a much higher frequency of negative events, compared to those with a significantly longer generation time. And to avoid unsustainable, exponential growth, the much faster branching generation sequences may even require to be stopped more often (see below). As a result it can be concluded that an increasing generation time demands a different niche situation for a sustainable multiplication. This in the sense that the niche conditions for multiplying forms with a longer generation time must be relatively more steady and reliable for the individual generation sequence concerned.

The reference to the generation time helps also to avoid confusion with the stability and reliability of environmental conditions as such. To give an example, certain prokaryotes live under conditions, such as hot springs, which are extremely reliable and stable in the sense that they exist probably unchanged since the beginning of evolution (see e.g. Knoll A.H. 2003). But this does not exclude that the individual generation sequence of these prokaryotes carries a high risk of being stopped.

The generation time was already connected to the relative complexity of a form. It was said that an increasing complexity results in more structural details and demands therefore more time to be reproduced. Thus, complexity increase during evolution should result in longer generation times. And indeed, to my knowledge a corresponding correlation is generally found in nature. As a consequence of what was said about niches, this suggests a direct link between the complexity of a form and its niche situation. To illustrate this important point let's consider the following. For sustainability reasons generation sequences need to be in balance with their environment. Consequently, as long as the environmental conditions are stable, the number of generation sequences should be stable too. At the same time generation sequences are persistently hit by negative events or changes. As long as the generation sequences can compensate the loss with sufficient branching events per time, they can avoid the risk that all get stopped. Obviously each branching corresponds to a multiplication event of the respective multiplying form. And because more complex forms multiply slower, their generation sequences show fewer branching per time. If the generation sequences of more complex forms are therefore hit by negative events with the same frequency as low complex forms, they are obviously at great risk to disappear.

In summary, a niche corresponds to an environmental situation that allows multiplying forms to adapt and continue their generation sequences. Here it is of importance that faster multiplying forms can tolerate more negative events per time, without all generation sequences getting stopped. In contrast, generation sequences of slower multiplying forms split up less frequently and tolerate therefore relatively less negative events per time, to stay in a lasting balance with the environment. And because more complex forms correspond in general to such relatively slower multiplying forms, it can be concluded that their individual generation sequence needs a relatively more stable and reliable niche situation. In the following I will deliver arguments as to why this interrelation between niche situation and complexity represents indeed a main principle of evolution.

Complexity increase is directed towards a mounting probability of multiplication

As next step on the search for a new understanding of complexity increase during evolution, the insights about niches can now be put into relation with the previously introduced probability of multiplication. As said, the latter captures the interdependence between environment and multiplication. It reflects the likelihood by which the individual generation sequence of a particular form is stopped by all kinds of causes, external and self-made. The higher the probability of multiplication, the lower is this risk for the individual generation sequence, relative to a longer time period encompassing multiple generations. And if the generation sequences of more complex forms show fewer branching per time, then they need obviously a relatively higher probability of multiplication. As a result it can be concluded that the complexity of a multiplying form and the probability of multiplication of its generation sequences are connected - this in the sense that the greater the complexity, the higher the probability of multiplication. Obviously this correlation fits well with our everyday experience. If something is more complicated and time consuming to produce, we are more attentive to ensure that it is not destroyed than is the case with something that can be produced fast and cheap.

The discovered correlation between complexity and probability of multiplication can be summarized in the following insight, which corresponds to a central dogma of *Interaction Theory*:

- The complexity increase of multiplying forms during evolution goes into the direction of a rising probability of multiplication of the generation sequences concerned.

This insight signifies that the individual generation sequence of a complex form has a higher average likelihood to continue over time than the individual sequence of one with a low complexity. This important correlation gives therefore the complexity of multiplying forms a clear qualitative meaning, which can be expressed as follows:

- The complexity of a biological form determines the average likelihood by which its individual generation sequence can continue over time under the respective niche conditions. This in the sense that the more complex a respective form is, the higher is this likelihood for the individual sequence.

Obviously this understanding about biological complexity is fundamentally different from established thinking. In addition, it provides the basis for the formulation of a law-like process for complexity increase during evolution, as will become clear in the following.

The new understanding of biological complexity has also obvious implications for our human self-concept and our role in biological evolution. If we accept *Homo sapiens* as the most complex living beings, at least on earth, it would mean that our emergence during evolution was the result of a process directed towards a rising probability of multiplication. In other words, complexity increase during evolution would have a direction. This stands in sharp contrast with evolutionary orthodoxy, which sees biological complexity and forms such as *Homo sapiens* as a purely accidental outcome (for a critical discussion of evolutionary orthodoxy see e.g. Conway Morris S. 2003). On the following pages and in particular in the second part of the book I will deliver arguments as to why biological complexity follows indeed a directed process towards a rising probability of multiplication. On the other hand it will also become clear that a rising probability of multiplication as such is necessary but not sufficient for complexity increase during evolution. First however, I will discuss the different niche requirements of multiplying forms in more detail.

1.5. The two fundamental types of niche

After having mentioned already the influence of negative environmental changes or events on the multiplication, I would now like to consider also the self-made part. This in the sense that generation sequences can also become stopped, because the multiplication itself can change the environment negatively. What can be said about niches in this respect?

Environmental conditions can obviously become exhausted or even destroyed by exponential growth. This means that either the resources become exhausted by over-consumption or the multiplication destroys the niche conditions in another way. In the search for environmental conditions which can serve as a niche, it is therefore necessary to consider the potential impact of growth on the niche factors. Here, a fundamental dilemma becomes obvious. If on the one hand life means multiplication, then multiplying forms will show an inevitable trend for self-enhancing, exponential growth under favourable conditions. And a niche provides obviously such favourable conditions. If on the other hand however, this growth will irreversibly change or destroy the very same conditions, the corresponding environmental situation can in principle not serve as a niche. There are two solutions to overcome this dilemma; either the niche conditions tolerate self-enhancing growth, or the multiplication occurs in a way that it does not exhaust or destroy the niche. This results in the following two fundamental types of niche:

1. Regenerating niches resistant to growth, characterized by environmental conditions that regenerate after their exhaustion or destruction by growth.
2. Niches at risk of becoming irreversibly exhausted or destroyed by growth, demanding a kind of multiplication that avoids unsustainable growth.

In consequence, for generation sequences to continue over time, they need to occupy one of the two fundamental types of niche. Transferring this insight into biological reality does mean that biological forms should be divisible into two fundamental groups, depending on their adaptation to the one or the other niche situation. Obviously these two groups should become visible during evolution, what is indeed the case – see below.

In the following I will describe the two fundamental types of niche and how they allow generation sequences to continue over time. I will also show that the split of organisms into sexually and non-sexually reproducing forms can be linked with the corresponding adaptation to the two types of niche.

Regenerating niches resistant to growth

This niche type is based on the following principle. A form is exposed to its niche conditions and starts to multiply. The result is exponential growth causing the resource(s) to become exhausted so that the multiplication finally stops. If this happens the forms may react with the transformation into a robust, inactive state, as a way to increase the chance to survive the period with negative environmental conditions. Over time the niche resource(s) regenerate and those forms which have survived can restart to multiply. The result is an oscillation between periods with exponential growth and stagnation, with the number of forms showing periodic peaks, while their average number can stay more or less constant over time. The time between two peaks would depend on how long the niche conditions need to regenerate. In summary, regenerating niches resistant to growth can provide a positive probability of multiplication to the generation sequences of forms with the ability to endure the negative interim periods. The adaptation to this type of niche corresponds therefore to one possibility of how generation sequences can proceed over deep time!

Asexual micro-organisms and in particular prokaryotes show a growth behaviour that follows the principles of regenerating niches. In the case the environmental conditions are favourable, their multiplication results in exponential growth and after the resources are exhausted they transform into robust endospores (Cooper S. 2012). Clearly the adaptation to this type of niche depends on the capability to use resources which can regenerate after exhaustion. If biological resources are ignored for the moment, it can be expected that resources with the capability to regenerate easily after exhaustion are normally relatively simple substances or factors, which are produced by permanent or regular reoccurring environmental processes. H_2S from geothermal origin that is used by certain micro-organisms as energy source is an example; sunlight and the use of CO_2 as carbon source is another. The ability to use relatively simple molecules, such as H_2S and CO_2, as well as inexhaustible factors such as sunlight, needs however exceptional biochemical capabilities. Therefore it is to be expected that forms adapted to this type of niche should in general be characterized by exceptional biochemical power. And actually this is precisely the case: micro-organisms in general and prokaryotes in particular are known for their strong biochemical capabilities (see e.g. Margulis L. 1993). This raises the suspicion that more complex, sexually reproducing forms may be adapted to a different type of niche, which turns out to be indeed the case.

The mentioned qualities of growth-resistant niches allow an interesting assumption regarding the beginning of evolution. If the generation sequences of life started with the multiplication of relatively simple molecules, as it is generally believed, it cannot have happened in this type of niche. The reason is that it is extremely unlikely that respective, relatively simple molecules would have had the necessary catalytic power. Later I will come back to this important insight.

With growth-resistant niches we enter a world under the reign of the so-called Red Queen (for the Red Queen hypothesis see Van Valen L. 1973). In brief this means that the respective forms are exclusively exposed to selection on fitness and have therefore to multiply as fast as possible. The result is exponential growth whenever possible. In other words, the competition for multiplication success under an exclusive selection of fitness causes a general growth contest. This means, all forms are multiplying and those multiplying faster will finally supersede the others. And unsurprisingly, such a general growth contest can only be sustainable in growth-resistant niches.

The idea of a general growth contest has strongly marked traditional evolutionary thinking. In the following I will show that this is, however, only part of the story. Multiplying forms can also adapt to the second fundamental type of niche, in which the competition for multiplication success occurs in a different way than via a general growth contest.

1.6. The role and significance of reproductive interactions

The second fundamental type of niche, allowing generation sequences to proceed over deep time, is again environmental conditions that provide sufficient stability and reliability for corresponding generation sequences. In contrast to the first type of niche however, they are at risk of becoming irreversibly exhausted or destroyed by exponential growth. Thus, the respective resources and/or conditions cannot regenerate in the same way as those of growth-resistant niches. To be sustainable, the multiplication must therefore avoid a general growth. This requires what I call reproductive interactions.

In the second fundamental type of niche the resources or conditions can also regenerate, but only within certain limits. Thus, an over-consumption puts them at risk of becoming irreversibly exhausted or destroyed. A concrete example is the prey for a predator; after the last prey is hunted down, this 'resource' is irreversibly gone. However, the negative consequences of growth may not only concern respective resources. It can also concern other niche factors, which may become irreversibly destroyed or changed as a direct or indirect consequence of unsustainable growth. And obviously in a situation in which the niche conditions are definitively gone, the transformation into a robust, inactive state would no longer be helpful. Hence, in niches at risk of becoming irreversibly exhausted or destroyed by growth, the necessary competition for multiplication success between the individual forms must occur in a different way than the general growth contest in growth-resistant niches - otherwise, the respective generation sequences risk being stopped by what I call self-made causes.

Now, how is a much more balanced multiplication possible that avoids the roller-coaster-like growth pattern of a general growth contest? The answer is that it becomes possible by a mutual dependence of different kinds of forms with regard to their multiplication. This means one form cannot multiply without the other, and vice versa. The term reproductive interaction stands for such a mutual dependence. As the reader will see, this type of interaction is not only essential for generation sequences to proceed in the second fundamental type of niche, but also of central importance for biological complexity increase. In illustration 1 the basic principles of a reproductive interaction at molecular level are shown.

Illustration 1: The basic principles of a reproductive molecular interaction

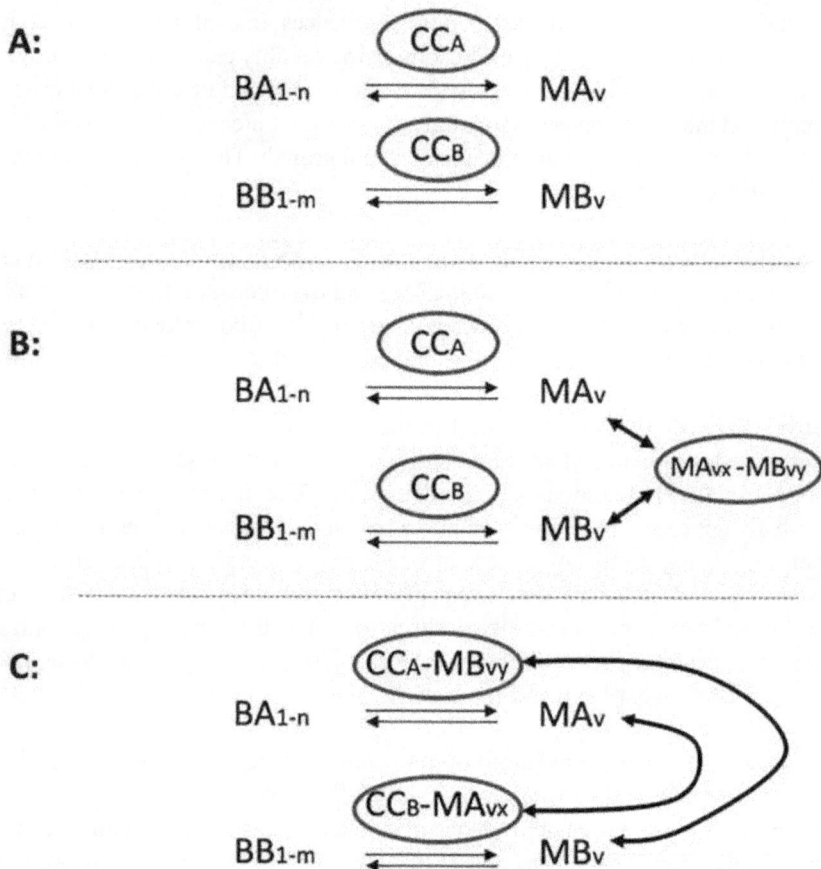

Simplified interaction situation with two molecular forms MA and MB. The molecular forms could be polymer-like molecules that are formed by respective catalytic complexes from a number of building blocks BA_{1-n} or BB_{1-m} respectively. The following 3 situations can be differentiated with regard to interactions:

A. MA and MB are formed by catalytic complexes in the presence of their building blocks BA, or BB respectively. The formation leads to MA and MB variants that differ in the order and number of the linked building blocks (MAv and MBv). Due to a limited supply of building blocks, the reaction results in a chemical equilibrium situation between formation and degradation of MAv and MBv, respectively.

B. Certain variants of the molecular forms (MAvx and Mbvy) can form a complex via reversible molecular interactions, such as hydrogen bonds and/or ionic bonds. As a result, the correspondingly interacting variants are deprived of the balance between their formation and their degradation and thus enriched relative to other variants. In other words, they are multiplied due to the interaction. If a new variant of MA or MB shows better interaction qualities, i.e. a higher interaction competence, it can displace other variants from the complex. This situation would correspond to a simple, non-reproductive molecular interaction.

C. MAvx is part of the catalytic complex CC_B, responsible for the formation of MBv, and MBvy is in turn part of CC_A, responsible for the formation of MAv. As in example B, new, better interacting variants can displace other variants from the complex and therefore accumulate. The variants that are part of the catalytic complex must support the catalytic reaction and thus indirectly their own formation, since the latter would otherwise come to a stop. This situation would correspond to a simple reproductive molecular interaction.

Instead of two partners, the interaction could also comprise more molecular forms, according to a chemical network.

Legend illustration 1:

- MA / MB = Molecular forms A and B, respectively, consisting of n or m different kinds of building blocks

- BA_{1-n} / BB_{1-m} = n or m different kind of building blocks of MA and MB, respectively

- MA_v / MB_v = Variants of MA or MB that differ in the order and number of their respective building blocks and thus in their specific interaction ability

- MAvx /MBvy = Specific variants of MA or MB with strong molecular interaction qualities

- CC_A / CC_B = Catalytic complex for the formation of MA and MB, respectively, if BA_{1-n} or BB_{1-m} are available

- CC_A-MBvy / CC_B-MAvx = Catalytic complex for the formation of MAv and MBv, respectively, with MAvx or MBvy being part of it

Reproductive interactions means simply that one form cannot multiply alone, but rather needs at least one interaction partner for a joint multiplication. To illustrate this, let's first look at molecules with the capability to kick off generation sequences by multiplying, i.e. to make more of itself. Under favourable conditions this type of multiplication will inevitably result in exponential growth and, in a world with exhaustible resources, finally end in an equilibrium state between formation and degradation of the corresponding molecules. Consequently, the multiplication will stop and so the generation sequences too. In the case of a reproductive interaction the situation is different. An overall equilibrium situation does not automatically stop a 'making more' of certain variants (see illustration 1). A new, better interacting variant can replace other variants in the catalytic complex, which is at the centre of the molecular interaction in illustration 1. This means that in each generation, better interacting variants can accumulate in the catalytic complex, where they influence the making of new forms. For a respective interaction between protein and polynucleotide for example, it would depend on the specific amino acid sequence of a protein variant, how well it can interact with a particular polynucleotide (see e.g. Maynard Smith J. and Szathmáry E. 1995). In the moment a new protein variant appears, which interacts better with the polynucleotide partners, it can replace other protein variants in the respective complex. If this variant can further influence the amino acid sequence of newly made proteins in its own sense, it will cause corresponding generation sequences. And because this influence cannot be absolutely accurate, new variants with different interaction qualities will continue to emerge. Hence, reproductive interactions allow a kind of selective multiplication of better interacting variants, even if the overall protein formation has reached an equilibrium state. The base would be the mutual selection between the interaction partners, regarding structural qualities that are relevant for their molecular interaction. This would be fundamentally different from a general growth contest driven by selection on fitness, as in the cited experiment by Spiegelman (1970). In summary, the principle that better interacting variants become multiplied while the total number of forms can stay in a balance corresponds to a solution as to how generation sequences can occupy niches at risk of becoming irreversibly exhausted or destroyed by a general growth contest.

That corresponding molecular interactions are of significance in biology is documented by the fact that the molecular basis of cells is marked by a reproductive interaction between DNA, RNA and protein. These fundamental molecules of life depend on each other, in order to become multiplied – i.e. RNA needs DNA and protein, DNA needs RNA and protein, and protein needs both kinds of nucleic acids (Campbell N.A. et al. 2008). In the light of what was said about the resource situation in the early molecular phase of evolution, this mutual dependence would reflect the need for reproductive interactions at the beginning of all generation sequences. In other words, the mutual dependence between DNA, RNA and protein, which represents the basis of the cellular metabolism, would be the visible documentation that evolution had to start with reproductive interactions between different kinds of molecules. This would have been necessary so that the generation sequences of life not become immediately trapped in a thermodynamic equilibrium situation.

The two types of organisms

Reproductive interactions can in principle occur between all sorts of multiplying forms, be it between different kinds of molecules, cells or organisms. In the following, I want to expand the discussion therefore beyond molecular forms. From the *Interaction Theory* point of view, the decisive aspect of reproductive interactions is that the multiplication success of a biological form depends on what can be called its interaction competence. Consequently, only variants with a sufficient interaction competence can multiply, and others not. Reproductive interactions are therefore the means by which organisms can influence their population growth. And this in turn has the consequence that reproductive interactions are mandatory for organisms, in order to occupy sustainably those niches that are irreversibly exhaustible or destroyable by exponential growth. In the following I will deliver arguments as to why precisely this is the case. First, however, I will consider whether it is visible in nature that biological forms are adapted either to the one or the other fundamental type of niche.

The introduction of the two fundamental kinds of niche gives a new meaning to the differentiation between forms that multiply either alone or via a reproductive interaction. Solitary multiplying forms would be able to cause generation sequences proceeding over geological periods of time, if they can use conditions which regenerate after an exhaustion by growth. The competition between the different variants would be driven by selection on fitness, favouring those with a higher number of descendants (Msuc = N / t (EC)). Consequently, solitary multiplying forms are exposed to a general growth contest, where all forms multiply and those multiplying faster will eventually prevail. Under favourable niche conditions, this contest causes an exponential growth and results eventually in the exhaustion of resources. In contrast, interacting forms would depend on each other to become multiplied. This allows a selective 'making more' of particular variants, without automatically resulting in overall growth. The generation sequences of interacting forms would therefore be able to use niches at risk of becoming irreversibly exhausted or destroyed by growth. In summary it can be concluded that solitary multiplying forms are trapped in a general growth contest that pro-

41

vokes exponential growth whenever possible. In comparison, reproductive interactions can result in a selective 'making more' of certain variants and without the necessity for overall growth.

The consequences of these conclusions are that living forms must belong either to the interacting or the solitary group, with each of them adapted to its respective kind of niche. And indeed, it is possible to divide biological forms into two clearly distinct groups. They belong either to the asexually or sexually reproducing organisms. And obviously, the asexual group corresponds to the solitary multiplying forms, of which microorganisms, with their rollercoaster-like growth pattern, represent a prototype (for bacterial growth see e.g. Cooper S. 2012). In this context it is striking that the forms in this group are in general of relatively lower complexity.

On the other side, the evolution of life has also produced animals and plants of higher complexity. These forms show normally a more balanced and less rollercoaster-like growth pattern. Following the previous arguments, the reason would be that their niche conditions are at risk of becoming irreversibly exhausted or destroyed by exponential growth. Consequently these organisms would need reproductive interactions for their sustainable multiplication. And actually, most of the higher organisms depend on sexual reproduction with two genders, which corresponds to the main form of reproductive interaction during evolution (see below). In addition, if the niche situation dictates the need for sexual reproduction, then heterotrophic forms should in general be more concerned than autotrophic, because the latter can use sunlight, which corresponds to a non-exhaustible energy source. And indeed, this is in line with the fact that asexuality is in general more frequent with plants than with animals (see e.g. Wesson R. 1991).

In summary, the discussion of how generation sequences can continue over deep time resulted so far in the identification of two fundamental types of niche, in which biological forms can either multiply alone or via interaction. In accordance with this demand living forms can be divided into two distinctive groups, the asexually and the sexually reproducing organisms. In addition, observable differences in the growth behaviour of the two groups support the argument that they are adapted to the two fundamental types of niche. Last but not least, it is striking that interacting forms are generally more complex than solitary multiplying forms.

Interaction competence and multiplication

After having introduced reproductive interactions with the example of molecular forms, I will now discuss reproductive interactions between organisms in more detail.

Generally speaking, reproductive interactions have the consequence that the multiplication success of the individual form becomes dependent on its specific competence to gain interaction partner(s) for its multiplication. Thus it demands additional qualities, which can be described as interaction competence (IC). This term serves in the following as a measure for the relative prospect of success of an individual variant of a certain kind of multiplying form, with regard to the competition for interaction partners. And because the individual interaction competence influences the multiplication success, it results in a specific type of selection, which I will call mutual selection on interaction competence or simply mutual selection. This type of selection occurs in each generation. It is driven by differences between the coexisting variants with regard to interaction relevant qualities. In the case of interacting molecules, such as protein and RNA, the differences in interaction relevant qualities would be the consequences of variations in the amino acid or nucleotide sequence. In the case of organisms, the reason would be the genetic variation between the individuals competing for interaction partners. The influence of the individual interaction competence can be expressed as follows:

$$Msuc = IC \times N / t \ (EC)$$

Here, the interaction competence IC can vary between zero and 1. It is 1 for a variant with the highest interaction competence in its particular interaction community (see below). As a result, this variant can have in principle its maximum possible number of descendants per time (N/t) under the given environmental situation (EC). In other words, it can fully exploit its multiplication potential. In the case where a variant is not able to interact at all, its IC would be 0 and hereby its multiplication success too. In general, the interaction competence of individuals would vary between both extremes, depending on their respective competitiveness in gaining an interaction partner. Consequently the multiplication success of interacting forms is influenced by two aspects. One would be the interaction-related part, expressed by the term IC. The second would be the fitness-related part N / t (EC), corresponding to the maximum possible number of descendants, if IC were 1. For the following discussion it will be central that the interaction part remains independent of the fitness part, in the sense that a high interaction competence does not automatically correlate with a high fitness and vice versa.

Now, why is it important that the multiplication success is influenced by the interaction and the fitness part independently? For an answer, let's consider the example of organisms with sexual reproduction, such as a peafowl. As sexually reproducing organisms the female and male birds need each other to multiply. The result is a mutual selection on interaction competence, which in this case corresponds to sexual selection. The result of this selection is very well visible in the appearance and courting behaviour of the peacocks (for sexual selection see e.g. Halliday T.R. 1994). In line with above argumentation, the individual variations in their plumage and mating behaviour would document the differences in interaction competence between the cocks. And obviously these qualities influence the multiplication success, however not only via female choice, but also via the individual fitness, in the sense that the more eyecatching peacocks carry a higher risk of being spotted by predators. This shows nicely the central point about reproductive interactions, namely that a high interaction competence does not automatically make the individual also fitter. It is this independence that corresponds to a kind of entry ticket for niches irreversibly exhaustible or destroyable by growth. The independence makes it possible that mutual selection can build a barrier against a development towards a general growth contest and the associated risk of exhausting or destroying the niche by unsustainable growth.

This reasoning stands obviously in sharp contrast to the widespread view that sexual attractiveness is simply the visible proof for genes that guarantee health and fitness (for a critical discussion of this view see e.g. Goodwin B. 1994 or Rose S. 1997). Here I must admit that the naive underlying connection between attractiveness and fitness astonishes me. It is better not to imagine the consequences if this logic were to be applied to human society. I'm confident that the arguments outlined in the following will convince the reader that this is not the case and that on the contrary, it is fundamental for the evolution of life to keep the fitness and the interaction competence of individuals separate.

If the above considerations are correct, it should be expected that a reproductive interaction, such as sexual reproduction, does actually possess mechanisms to maintain the independence of interaction competence and fitness. And indeed, a major characteristic associated with sexual reproduction is the recombination of homologous chromosomes during meiosis (see Maynard Smith J. and Szathmáry E. 1995). The recombination mechanism does exactly what is expected: it acts against a permanent coupling of genes, respectively alleles to be precise. Via the recombination of homologous chromosomes, alleles can potentially become separated in each generation. This means that if by chance an individual form emerges which combines high fitness with high interaction competence, thanks to genetic recombination, its descendants may no longer possess this combination. From the viewpoint of *Interaction Theory* the genetic recombination is therefore a necessity that preserves the independence of interaction competence and fitness for the sustainable occupation of irreversibly exhaustible niches. In contrast, the benefits of recombination are difficult to recognize under the sole aspect of fitness (see Eldredge N. 1995 or Maynard Smith J. and Szathmáry E. 1995).

At this point a fundamental remark about sexual reproduction seems necessary. This type of reproduction and the associated sexual selection is well documented and part of common evolutionary theory. The popular explanation of why this mode of reproduction has developed during evolution is, however, controversial (see e.g. Eldredge N. 1995, Goodwin B. 1994 or Rose S. 1997). I will not enter further into this discussion, because it would distract from the principal subject, which is the complexity increase during evolution. Important in this regard is however, that from the point of view of *Interaction Theory* sexual reproduction is without doubt a very central type of reproductive interaction, but not the only one. Evolution resulted in a mutual dependence between different types of multiplying forms. The probably most striking example on the molecular level is the already mentioned mutual dependence between DNA, RNA and protein. On the level of organisms it is clearly sexual reproduction. Nevertheless, other types of interaction between organisms are important too. Examples are flowering plants and pollinating insects, social interactions within species such as humans, dolphins or chimpanzees, but also the mutual dependencies between the different castes of eusocial insects. In all cases, the multiplication success is influenced by a need for interaction competence. A key message of this book is therefore to see the different kinds of reproductive interactions under the same aspect: they exist because the mutual dependence between the interaction partners results in mutual selection, which can lower the risk of irreversibly exhausting or destroying the respective niches by unsustainable growth. For this reason, I consider here not only sexual reproduction, but reproductive interactions in general. And for the same reason, I'm focusing not specifically on sexual selection but on mutual selection in general.

Mutual selection and interaction communities

Reproductive interaction stands for a situation in which at least two sorts of forms need each other for their multiplication. Obviously this corresponds to some kind of symbiosis. To differentiate it from other kinds of symbiotic relations, the term reproductive interaction will therefore be used if two criteria are given. Firstly it must be a mutual dependence in which the interaction partners cannot multiply alone and need each other for their multiplication. Secondly, the interaction partners are exposed to a mutual selection on their interaction competence within an interaction community. In the following I will discuss the mutual selection within interaction communities in more detail.

Decisive for the precise nature of mutual selection is the relevant interaction community, which means the distinct group of forms that are actually interacting with each other. The relevance of the interaction community for evolution is given by the fact that each individual can only interact with a limited amount of other individuals. For example, a male lion in Africa is not in competition for females with all other African male lions. He is only in competition with those he will actually be confronted with during his lifetime. Therefore the impact of the interaction competence on the multiplication success depends always and exclusively on the relevant interaction community.

The demand that mutual selection is determined only by the respective interaction community causes a key difference to selection on fitness. For selection on fitness it can be assumed that something like fitness peaks exist, which depend on the particular niche situation (see e.g. Maynard Smith J. and Szathmáry E. 1995). These fitness peaks give selection on fitness a direction. For example, in the previously cited Spiegelman trial the relevant fitness peak corresponds to the shortest possible nucleotide-sequence allowing the fastest possible multiplication. In contrast, comparable interaction peaks do not exist. The reason is that as long as both types of selection stay independent, mutual selection depends solely on the selection criteria that occur in the respective interaction community. And the precise criteria for this mutual selection would obviously be the consequence of the prevailing preference, which again depends on the particular variant composition of the interaction community. Thus, if the variant composition of interaction communities is accidental, it might result in differences regarding the precise mutual selection criteria. Certain aspects may be more important in one community than in another, depending on the respective members and their specific preferences. This means that the precise nature of mutual selection would be unpredictable, because the profile of the variants in an interaction community can change by accident.

On the genetic level, the variant composition of an interaction community would depend on mechanisms such as mutation and genetic drift. The resulting genetic variation would determine the respective interaction competence of the individuals as well as the prevailing preference for interaction partners. And because mechanisms such as mutation or genetic drift are accidental, their consequences for the precise nature of mutual selection within an interaction community are unpredictable. Here however, it can be expected that in smaller, isolated interaction communities the criteria for mutual selection can change more easily over a few generations than in large communities. The reason lies in the fact that mutations and genetic drift have a stronger impact on smaller groups (for mutation and genetic drift in evolution see e.g. Stearns S.C. and Hoekstra R.F. 2005). Consequently, if a big interaction community splits up in smaller and isolated communities, those may start to develop different criteria for interaction competence. As I will discuss in a later section, corresponding changes in mutual selection can be linked to the evolution of species, and the question of why new species emerge.

Natural selection versus dual selection

In the postulation of two independent kinds of selection lies a difference with traditional Darwinism, for which natural selection is the decisive factor in evolution (Darwin C. 1859). *Interaction Theory* sees the need for two independent forms of selection, because a part of the generation sequences of life depends on niches irreversibly exhaustible or destroyable by growth. The difference between both types of selection is clearly visible with regard to their respective influence on the introduced probability of multiplication. As said, the latter reflects the likelihood by which the individual generation sequence gets stopped by external or self-made causes. This likelihood is influenced by an exclusive selection on fitness as follows. If a solitary multiplying form is exposed to favourable environmental conditions, selection on fitness promotes those variants which are relatively fitter, until a point where a fitness peak is reached (see again e.g. Maynard Smith J. and Szathmáry E. 1995). And increased fitness means that either the generation sequences are stopped less frequently by external causes or they show more branching events per time or both. At the end this results under favourable environmental conditions in exponential growth which will cause self-made changes, in the sense that the niche situation becomes exhausted or destroyed. In the case this exhaustion or destruction will be irreversible and all generation sequences will come to an end, the probability of multiplication would be zero. This leads to the insight that an exclusive selection on fitness in a niche irreversibly exhaustible or destroyable by growth would not allow a sufficient probability of multiplication for generation sequences. Thus, to continue over time and to conserve a positive probability of multiplication, generation sequences have to stay in a balance with the environment and to avoid an exclusive selection on fitness endangering this balance.

Compared to this, the mutual selection associated with reproductive interactions can have a different influence on the probability of multiplication. It can prevent a development towards exponential growth in irreversibly exhaustible or destroyable niches. For example, let's imagine an interaction community with a sufficient probability of multiplication, which multiplies under stable environmental conditions. To keep the balance with its environment, the community needs to avoid a development that results over time in an overall increasing number of generation sequences and the associated risk of unsustainable growth. That can happen if new variants, with on average more descendants per time, become progressively dominant. In an interaction community however, a new variant can only supersede other variants, if it also shows over the generations a sufficient interaction competence. If this is not the case, mutual selection can make it difficult for the fitter variants to prevail and the overall balance between multiplication and environment can be conserved. In other words, mutual selection can build a barrier against a selection on fitness favoured development towards more progenies per time. As said, this requires that the interaction community be adapted to a relatively stable environmental situation and that the qualities determining the individual interaction competence are not tightly linked with those determining the individual fitness.

To illustrate this important point I would like to use the example of a bird species, of which the females lay in average 2 eggs per year. This reproduction rate allows the birds a sustainable multiplication in their respective niche and gives their generation sequences a sufficient probability of multiplication. New variants with the capability to produce on average 3 eggs would occasionally appear. If they were to become dominant, the sustainable balance between environment and multiplication may come in danger. The new variants can, however, only supersede the others if they own a sufficiently high interaction competence. But if the qualities, respectively the genes, which are responsible for the numbers of eggs, are not coupled with those that are responsible for the individual's mating success, the chance is low for the 3-egg variant to become dominant. And even if both traits, the capability of laying 3 eggs and a high interaction competence, become accidentally combined in an individual, genetic recombination can re-separate them in one of the following generations. In the case that each of the traits is the result of a combination of multiple genes (as it is probably mostly the case, see e.g. Goodwin B. 1994), then the barrier becomes even higher.

To avoid any misunderstanding, the above should not be misunderstood in the sense that mutual selection on interaction competence prohibits any increase in fitness. This would be wrong and can clearly not be the case. The exact consequences of a dual selection on both interaction competence and fitness would rather strongly depend on the respective environmental situation. This means that if the generation sequences of a form are in a stable balance with their environment, it is obviously essential for the multiplication success of an individual to own a high interaction competence. Consequently, an interaction community with an adequate mutual selection would be able to preserve the balance with the environment, by avoiding strong population growth.

The situation changes however, if the interaction community experiences whatever kind of challenge for its survival. Now the capability to deal with the respective challenge becomes decisive for the individual survival and herewith the faith of the respective generation sequence. An example would be a predator species that is suddenly at risk of becoming extinct, due to new, well protected forms of prey. In such a situation, new predator variants with an improved hunting success would have a strong benefit, which could compensate an average or even low interaction competence. As a consequence, the variants with a correspondingly increased fitness can become dominant, and a new balance with the environment becomes possible. In summary, selection on fitness would be essential when the survival of an interaction community is challenged, while mutual selection would be essential when the corresponding interaction community is in a stable balance with its environment.

The fate of interaction communities

In this section I would like to discuss the fate of interaction communities in irreversibly exhaustible niches. For this, imagine an interacting form, such as a sexually reproducing species adapted to particular niche conditions, which can be found in multiple geographies. Given the wide distribution of the niche it can be assumed that the corresponding form is split into different, independent interaction communities, simply by geographical separation. The resulting communities would correspond to flexible entities, which may either exchange individuals regularly or sporadically with other communities, or become entirely isolated. Each of these communities can be seen as responsible for the fate of its generation sequences. And because of the dependence on an irreversibly exhaustible niche situation, the communities should stay in a sustainable balance between multiplication and environment. As said, the way to do so would be the dual selection on fitness and interaction competence. On the other hand it was also said that the precise articulation of mutual selection may start to vary if interaction communities become isolated. As a result, the self-made, mutual selection in separated interaction communities may become different, if they are separated long enough. And this in turn may have the consequence that some communities may no longer be able to keep a sustainable balance with their direct environment and eventually become extinct. On the other hand, for the long-term fate of a particular form it is not essential that each single of its interaction communities survives. To avoid extinction, it would be enough that at least some show a mutual selection, so that its generation sequences can continue.

Let's now turn to the question of how interaction communities can show an adequate mutual selection. In this respect it can be assumed that in certain niches the self-made selection needs to be strong and frequent, while in others weaker and less frequent. In sexually reproducing species this would correspond either to a strong sexual selection during each mating season or a situation with males and females forming a permanent, lifelong couple. But how can a form 'know' which type of mutual selection it needs? Here it is necessary to consider that mutual selection has no direct benefit for the interaction community in the present. Its potential benefit lies in the future and interaction communities cannot adapt their mutual selection to something that is only relevant some or even many generations later (for respective arguments see e.g. Williams G.C. 1966). In addition, I came a few pages earlier to the conclusion that the precise articulation of mutual selection within a particular community is unpredictable. The answer to above question of how interaction communities can show an adequate mutual selection can therefore only be that it is the environment that is decisive. In other words, the respective environmental conditions determine whether the generation sequences of a particular community with its mutual selection will survive or not. As said, here it would not be necessary that all interaction communities of a form show at any time a sustainable balance with their direct environment. For the corresponding generation sequences to continue, it would be enough that at least some, or even only one, can survive.

In summary, it has been concluded that the mutual selection in interaction communities builds a barrier against a development towards unsustainable growth. However, multiplying forms cannot know whether their respective self-made selection is adequate to sustain a balance between their particular environment and their multiplication. Therefore it can only be the environment that determines whether the generation sequences of a particular interaction community can continue or will end. Thus, the interaction community is a central entity of evolution from the *Interaction Theory* point of view.

1.7. Symbiosis and generation sequences

Following the discussion of reproductive interactions and niches irreversibly exhaustible or destroyable by growth, I will now briefly discuss symbiosis in a wider sense and its potential impact on generation sequences. The reason for this is that symbiotic relations can be seen as important with regard to complexity increase of solitary multiplying forms.

Reproductive interactions were defined as a mutual dependence between different types of forms, so that one type of interaction partner cannot multiply without the other. By comparison, I want to use the term symbiosis or symbiotic interaction for situations in which an independent multiplication of the partners is still possible. Obviously, the borderline between symbiosis and interaction is fluid. As an example, the relation between flowering plants and pollinating insects would correspond to a reproductive interaction from the point of view of flowering plants. For most insects however, it would be more a symbiotic kind of interaction, because they do not directly need the plants for the multiplication act as such. On the other hand, the plants are essential niche factors for the insects and therefore indispensable for their survival. Thus it is justified to say that the generation sequences of both partners need each other to continue over time.

Symbiotic and reproductive interactions also have in common that they provoke a mutual selection, if the respective interaction or symbiosis partners can choose between different variants. On the other hand, a particular case is given if originally independent symbiosis partners become united in a new entity or structure. Now the interaction partner is given and can no longer be chosen. The so-called endo-symbiosis between proto-eukaryotic cells and the precursors of mitochondria and chloroplasts is a concrete example, to which I will come back (for endo-symbiosis see Margulis L. 1993).

In summary, multiplying forms can interact either via a reproductive interaction, thus between different forms that are not capable of multiplying alone. Or they can interact via symbiotic interaction, which describes a situation in which the partners might still multiply independently. This differentiation between interaction and symbiosis is of relevance for the niche model introduced. While reproductive interactions would exclusively occur in niches irreversibly exhaustible or destroyable by growth, the symbiotic relations would not be restricted to a particular kind of niche (see below).

Restrictions for the probability of multiplication of solitary multiplying forms

A starting assumption of this book is the inevitable interdependence between living forms and their environment. Not only is the environment selecting the biological forms over the generations, but by their ongoing multiplication, the latter are also changing the environment. To acknowledge this mutual influence, the probability of multiplication was introduced as an indicator for a sustainable balance between multiplication and environment. In this context it was concluded that in niches at risk of becoming irreversibly exhausted or destroyed by growth, reproductive interactions are necessary to preserve a positive probability of multiplication. Now I will discuss how solitary multiplying forms can influence their probability of multiplication.

As said, solitary multiplying forms are forced by selection on fitness into a general growth contest and show therefore a rollercoaster-like growth pattern. In the case that a solitary multiplying form is feeding on another solitary multiplying form, a situation which can be described as an arms race would be the result (as previously mentioned this is often called a Red Queen hypothesis proposed by Van Valen L. 1973). What would be the consequence of such an arms race? For this, let's imagine a respective hunter-prey relation. The hunter causes a selection of the prey, in the sense that those prey variants which are relatively better protected have an advantage. In the case the selective pressure caused by the hunter is sufficiently high it can result in the accumulation of these variants. In response, this would cause a counter-selection favouring hunter variants, which are able to hunt the now prevailing, better protected prey more efficiently. In summary, adaptations of solitary multiplying forms, such as a better protection against a hunter, can only be beneficial for the individual multiplication success over some generations. In the longer term this benefit would disappear again. This can also be expressed via the probability of multiplication as follows: whenever a solitary multiplying form is exploiting another solitary multiplying form, its probability of multiplication will be influenced by the unavoidable arms race between the two parties. In the given example this means that a greater hunting success with the prevailing prey variant can increase the fitness of the corresponding hunter variant short term, but not the probability of multiplication of the corresponding generation sequences in the longer term. At the end, the arms race can even be negative for the probability of multiplication, because a party may eventually lose out.

The consequence of what was said would be that hunter-prey or parasite-host-like relationships between solitary multiplying forms should face restrictions. This in the sense that such relationships can only be sustainable, if a better exploitation of one party by the other does not endanger the probability of multiplication. Obviously this is different from what was said about interacting forms, namely that they can protect their probability of multiplication via the dual selection on interaction competence and fitness. Now, if this is true, it can be expected that in nature hunter-prey or parasite-host relationships exclusively between solitary multiplying forms are somehow different than is the case with sexually reproducing forms. And indeed, to my knowledge this is the case.

Let's consider for example prokaryotes, the prototype of solitary multiplying forms. Normally we are used to predatory interactions, in which "big feeds on small". Obviously this makes sense, because a larger size can make it easier to overwhelm a prey. Now when it comes to bacterial predators however, it is rather the other way around, in the sense of "small feeds on big" (Jurkevitch E. 2007). This would support the above assumption that a solitary multiplying predator is in permanent danger of losing the mentioned arms race with its prey. For the bacterial predator cells this means that the bigger they are, the more time the multiplication needs, compared to a smaller prey. Hence, they should stay small compared to their prey, as is the case.

It is further known that bacterial predator cells, such as *Bdellovibrio*, can engage relatively fast in an arms race, resulting in more resistant prey and more aggressive predators (Jurkevitch E. 2007). Again this is different from what we are used to in prey-predator relations between sexually reproducing organisms, which are normally relatively stable (see e.g. Pianka E. R. 2000). Later, in the context of the evolution of eukaryotes, I will come back to this important point.

What can be said about the impact of symbiotic relationships on the probability of multiplication? Here the adaptation of the symbiosis partner can result in an increasing mutual benefit (Margulis L. 1993). This means that a mutual symbiotic adaptation in the here and now can positively impact the probability of getting the corresponding benefit in the future. Thus, it can be concluded that symbiosis in general, and in particular if it provides a strong mutual benefit, has a positive impact on the probability of multiplication of the respective generation sequences. Obviously, this should be the case for all kind of forms, because the type of niche would not be relevant in this regard. Consequently, symbiosis must play an important role during evolution. Neither solitary nor interacting forms should simply evolve side by side without any other relation than competing for resources and space. Instead they should start whenever possible symbiotic relations, by coupling their multiplication in a positive sense. This would not only provide a benefit in the present, but also a positive long-term effect on the probability of multiplication of their generation sequences. And indeed, in nature symbiosis plays a very important role in the relationship between living forms. Without going deeper into this subject, I want to mention in particular the importance of symbiotic relations within ecosystems (see Margulis L. 1993 and 1998).

In summary it can be said that for generation sequences and their probability of multiplication, not only reproductive interactions but also symbiotic interactions in general should play an important role during evolution. Both types of multiplying forms, solitary as well as interacting, should therefore engage in symbiotic interactions wherever possible. This would not only have a short-term benefit, but also a positive impact on the probability of multiplication of their generation sequences. Related to this insight, I will show in the following that symbiosis also plays a key role in the complexity increase of solitary multiplying forms.

1.8. Complexity increase and the probability of multiplication

On the previous pages, I have started to lay the basis for a new understanding of the complexity increase during evolution. Central in this regard is the focus on generation sequences from the sustainability point of view and the introduction of two fundamental types of niche. Based on this, I will now try to understand why and how the complexity of multiplying forms increases, while their generation sequences are travelling through deep time. This will lead to the central insight that biological complexity is not accidental and meaningless, but follows a law-like process of complexity increase towards a mounting probability of multiplication. To arrive at this conclusion, I will first discuss two questions which result from insight about a correlation between probability of multiplication and complexity, as argued some pages before. Firstly, what does a rising probability of multiplication mean for the environmental adaptation? And secondly, how can multiplying forms acquire a higher probability of multiplication? The answers to these questions will then serve as starting point for the formulation of the law-like process of complexity increase.

Niche requirements for a mounting probability of multiplication

The first of above questions can also be expressed as follows: are the niche requirements of complex forms, because of a higher probability of multiplication, different from those of less complex forms and if yes, how? Again the answer needs to consider that the number of generation sequences must be in a sustainable balance with the environment. To start with, I will summarize the insights already gained about the relation between probability of multiplication and niches.

Niches were defined as environmental conditions allowing multiplying forms to be part of generation sequences proceeding over deep time. For this reason it was concluded that environmental conditions can only serve as a niche for generation sequences, if the likelihood that negative events will stop all corresponding sequences is sufficiently low. As said, negative events stand for all possible things, such as being killed by a beast of prey, or dying of starvation, or having a fatal accident etc. While the exact nature of these events can vary, they all have in common that they either eliminate forms or make at least their further multiplication impossible. The likelihood with which this happens would be crucial for the probability of multiplication. The higher the chance to be stopped by negative events, the lower is the prospect for the individual generation sequence to continue over deep time. In other words, the probability of multiplication, as measure for the likelihood by which the individual generation sequence gets stopped, depends on the environmental conditions, or niche to which a form is adapted. As a consequence it can be deduced that a higher probability of multiplication demands a niche situation with a relatively higher stability or reliability for the individual generation sequence. And if more complex forms own a higher probability of multiplication then they need a respective niche situation, whereas lesser complex, faster multiplying forms can tolerate a niche situation under which they are exposed to relatively more negative events or changes per time. This allows a first answer to the above question: more complex forms require a niche situation that causes for the individual generation sequence a lower risk of getting stopped by negative events. And obviously, this relatively higher stability or reliability of the niche situation would be the reason for a higher probability of multiplication of more complex forms.

Now, what does this mean for the two fundamental types of niches? With regard to growth-resistant niches occupied by solitary multiplying forms the answer is relatively simple. The less these forms are affected by negative events, the faster will be their growth, which results eventually in a temporary exhaustion of the conditions. In other words, if exponential growth tolerating niches stop generation sequences with relatively less negative events per time, the result can be a faster self-made degeneration of the conditions. This makes it difficult to imagine how generation sequences can follow an ongoing development in this type of niche towards a mounting probability of multiplication as a condition for a higher complexity. And indeed, in line with this assumption is the fact that prokaryotes, which correspond to the prototype of biological form adapted to growth-resistant niches, remained over billions of years structurally relatively simple compared to eukaryotic organisms.

Let's now look at niches at risk of becoming irreversibly exhausted by growth. If generation sequences were so far in a sustainable balance with their environment and then become suddenly exposed to less negative events or changes, also in this case the overall number of generation sequences can increase. Thus, if in irreversibly exhaustible niches less generation sequences are stopped, the risk of an irreversible exhaustion or destruction of the niche conditions can potentially rise. On the other hand it was said that reproductive interactions are a means to reduce the risk of unsustainable growth in this niche type. Consequently, interaction communities should in principle be able to adapt in a sustainable way to a niche situation, which is relatively more stable or reliable for the individual generation sequence. Thus, a complexity increase towards a progressively rising probability of multiplication should be possible in irreversibly exhaustible niches, which are occupied by forms with respective interactions. In line with this assumption is the fact that complex organisms are indeed characterized by different reproductive interactions, such as sexual and social interaction in animals, or in the case of angiosperms their interaction with insects. In addition, it will turn out that complexity increase goes together with a visible intensification of mutual selection.

In summary, from the *Interaction Theory* point of view the following can be said about the niche requirements for a rising probability of multiplication during evolution. The probability of multiplication can continually rise, if the corresponding niche situation is progressively more stable or reliable for the individual generation sequence. However, niches tolerating exponential growth are bad candidates for a respective development, because more stable or reliable conditions would result in a faster exhaustion by growth. By comparison, niches at risk to become irreversibly exhausted are occupied by multiplying forms with reproductive interactions. As a consequence of what was said about the role of these interactions, they might allow the avoiding of unsustainable growth, as a result of progressively more stable or reliable niche situations. This means that if complexity increase goes towards a rising probability of multiplication, it should occur in niches irreversibly exhaustible by growth and by forms with reproductive interactions, just as is the case (see below).

Complexity increase and niche conditions

What else can be said about niche requirements that might be relevant for evolutionary complexity increase? For solitary multiplying forms, with strong exponential growth under favourable conditions, the niche specific resources and the possibility that they become exhausted by growth should be of particular importance. The resources need to tolerate that the respective forms are forced by selection on fitness into a general growth contest. The precondition to this type of niche would be the ability to use resources which are not irreversibly exhaustible by their consumption, such as sunlight. After forms with this capability have emerged, they may cause directly or indirectly the emergence of additional, exponential growth tolerating niches. The result would be autotrophic and heterotrophic forms building together ecological networks, as they are known from micro-organisms (see e.g. Knoll A.H. 2003 or Margulis L. 1993).

In comparison, irreversibly exhaustible niches consist of resources which can regenerate within certain limits, but are irreversibly exhaustible by over-consumption, such as the prey for a hunter. This was identified as the reason why reproductive interactions would be required in this type of niche. With their dual selection they allow a more balanced population dynamic. On the other hand the natural environment cannot be absolutely constant, but will inevitably show fluctuations (see e.g. Gribbin J. 2004). In consequence, also niche conditions cannot be absolutely constant and will also show fluctuations. For forms multiplying in niches at risk of becoming irreversibly exhausted this means that their resource situation would alternate between favourable and unfavourable for growth. The average time period during which the situation is favourable or not should depend on the particular niche. A reason for such fluctuations can be periodic climate changes, such as those caused by the famous El-Niño. For the probability of multiplication of generation sequences this kind of fluctuation should be fine, as long as it happens within certain limits and a sufficient number of sequences can carry on. Nevertheless it does also mean that the population density of a form should oscillate over the generations. In other words, periods with an overall growing number of generation sequences caused by a favourable resource situation should be separated by periods with spare resources and a decreasing number of generation sequences. And indeed, this oscillation corresponds to what can be found in nature (see e.g. Pianka E. R. 2000 for population dynamics of the Lotka-Volterrra model). Consequently, the average period of time during which the environment provides favourable conditions for growth would be important for the question of how the niches of complex forms differ from those of less complex ones.

The following example can illustrate this important point. A particular niche would be characterized by conditions which are favourable for growth over periods of about 20 years in average, interrupted by short difficult periods. This means that a form with a generation time of 1 year, which is adapted to these conditions, can grow, in principle, over 20 generations. Here it is good to recall that in the case the form would double in each generation, this could theoretically result in 2^{20} more forms in 20 years! If this strong growth has irreversible consequences, it means that this specific niche situation is not suited for the generation sequences of this fast multiplying form. On the other hand, the respective niche conditions may be well suited for more complex forms with significantly longer generation times, such as 5 years. In this case the average period with favourable growth concerns 4 generations only.

Now, are these conclusions in accordance with reality? If yes, then a visible consequence should be that forms at the low end of biological complexity need to multiply the fastest, because they would be adapted to niche conditions which can change frequently and fast for the individual form – fast means within hours, rather than months or years. Prokaryotes correspond to such forms and they are indeed characterized by an exceptionally high multiplication rate (Cooper S. 2012). Correspondingly, this link between complexity and niche conditions is also visible at the high end of biological complexity. For example, highly complex living forms, such as humans, multiply relatively slowly and are adapted to environmental conditions that are comparatively constant and reliable over long periods of time, such as decades – otherwise the particularly long childhood of *Homo sapiens* would be difficult to imagine.

As a result, the insights about niche conditions and complexity can be summarized as follows: Complexity increase during evolution occurs in irreversibly exhaustible niches, which become progressively more stable or reliable for the individual generation sequence. This would be the reason why the individual generation sequence of more complex forms has on average a greater likelihood to continue over time, as visible in their rising probability of multiplication. This important insight leads to the following assumption:

- The underlying cause for complexity increase during the evolution of life is that over time some generation sequences acquire the capability to benefit from a more stable or reliable niche situation.

As it will turn out, it is exactly this principle that lays the basis for a law-like process of complexity increase. And it is obvious that this link between complexity and niche situation represents a key difference between traditional evolutionary thinking and *Interaction Theory*.

Niche factors and ecological networks

With regard to the discovered relation between complexity and niche requirements, I will now explore the reason why the generation sequences of more complex forms can benefit from a relatively more stable or reliable niche situation. Two solutions are conceivable. Firstly the capability to deal with negative events could depend on the complexity. This means that not the niches as such would be decisive, but the fact that more complex forms would own qualities allowing them to better tolerate negative events, compared to forms of lower complexity. In other words, with increasing complexity the forms could become more robust. The alternative view is obviously that not the forms as such, but their niche situation, would be fundamental different. This means the niche conditions could be more stable and less challenging for the individual generation sequence of more complex forms. In this case the complexity of a form would be decisive, whether it can adapt to such niche conditions or not. In summary, the relation between complexity and niche situation could be explained in two ways: either the niches are changing, or the robustness of the forms.

None of these two explanations looks, however, very promising. So is it difficult to imagine that increasing complexity means greater robustness. Let's take again the example of prokaryotes. Whatever kind of catastrophe will happen to the earth in a hopefully very distant future, in the case some few eukaryotes will survive, we can be absolutely sure that many more prokaryotes will survive too. In other words, the forms at the low end of biological complexity own an outstanding robustness (see e.g. Knoll A.H. 2003). On the other hand, the fact that multiple forms of different complexity live often happily together in the same environment does also not endorse the alternative view that the niches as such are fundamentally different. So, what then could be the reason? To find an answer to this question, it turns out that it is necessary to understand the niche situation of multiplying forms better. And for this it is helpful to differentiate between two principal categories of niche factors, namely the biological and the non-biological. The latter consists of physical factors, such as sunlight or water, and is therefore normally the same for the forms living at the same place. Thus, it may not be straightforward to find here an explanation as to why niches and complexity are linked. Therefore I will first look at biological factors and later come back to the non-biological.

Biological niche factors are other forms which do not belong to the same type or species but influence the multiplication in the respective niche. This influence can obviously be positive or negative. Thus the multiplication of one form influences the generation sequences of the other, either positively or negatively. From the *Interaction Theory* point of view, biological niche factors mean therefore a positive or negative influence of other multiplying forms on the generation sequences. In reality, such kind of mutual influence of the multiplication between different forms represents very probably a central part of the niche conditions.

The previously discussed symbiotic relations correspond to a positive example. However, for an understanding about the relation between complexity and niche characteristics, the negative influence may be of particular interest. The reason is that a sudden self-amplified multiplication of one party will impact negatively the stability or reliability of the niche situation (see e.g. Van Valen L. 1973). This argument does not only concern heterotrophic forms with prey-predator or host-parasite relations, but also autotrophic, such as plants in competition for space, sunlight or in their struggle with herbivores.

Now, can biological niche factors be responsible for the relation between complexity and niche requirement? To answer this question let's imagine for example a prey-predator situation, in which a specific form of prey represents the decisive resource for the multiplication of a predator. In this situation, a predator with a high multiplication rate, because of a low complexity with short generation time, faces a fundamental problem if the prey is significantly more complex and therefore much slower multiplying. A too strong growth of the predator population risks making the prey extinct. This danger should be higher if the difference is very big, and correspondingly lower if the multiplication rate of the predator is similar to or even below that of the prey. From the sustainability point of view it should therefore be expected that in prey-predator relations the complexity difference between both parties plays an important role. The relation should be more stable or reliable if either the predator is in about the same range or more complex than its prey. In other words, biological niche factors with a negative influence are potentially sensitive regarding complexity differences between the concerned forms.

In connection with biological niche factors a further aspect needs to be considered. If different forms start to influence mutually their multiplication success in a positive and negative way, the result is an ecological network or ecosystem. Assuming that all niches have biological factors, then all niches would automatically be part of such networks. This means that niches do not exist in isolation. Rather, they are part of ecosystems, which are a result of the negative or positive mutual influence between the different multiplying forms. Hence, generation sequences have to fit to a given biological diversity, in the sense that the influence of other forms on them, and *vice versa*, must be sustainable. For the complexity increase during evolution this has the consequence that whenever new, more complex forms emerge, their generation sequences need to integrate in a sustainable way into the existing ecosystem. In other words, complexity increase requires a suitable ecological environment. Later I will show that the emergence of more complex forms and the emergence of new ecosystems are indeed intrinsically linked.

Parasites and their probability of multiplication

From what was said about niche requirements and complexity it follows that parasites carry a particular risk. Their multiplication has a direct negative influence on their hosts and they own normally a much lower complexity with a shorter generation time than these. Obviously this fact needs to be brought in accordance with what was said before, namely that the negative coupling of the multiplication between forms of very different complexity is subject to restrictions. Therefore I want to discuss how parasites fit into the demanded relation between niches and complexity.

Like for any other form, the generation sequences of parasites need a sustainable multiplication. For this reason they need to avoid the extinction of their hosts and thereby their own. But how is this possible, if parasites can multiply much faster than their hosts? To answer this question, I must anticipate a fundamental difference between prokaryotes and eukaryotes, which will be discussed later in more detail. This fundamental difference would be that eukaryotic generation sequences can conserve qualities without fitness benefit much better than is typical for prokaryotes. This attribute becomes visible in the significantly larger genome of eukaryotes compared to prokaryotes. As I will discuss later, the significantly larger genomes can be seen as a key innovation that made the emergence of eukaryotes possible. It allowed the occupation of a certain type of niche that demands the conservation of genes, even if those show no fitness benefit over the course of several generations. In comparison, prokaryotes would not be able to occupy this type of niche in a sustainable way, because they lose genes without a fitness benefit during some generations relatively fast, so that their genomes stay compact (for prokaryotic and eukaryotic genomes see e.g. Margulis L. 1993 or Maynard Smith J. and Szathmáry E. 1995).

From the *Interaction Theory* point of view, the capability of eukaryotes to preserve larger genomes would be decisive for host-parasite relations. It would allow the eukaryotic hosts to keep much less complex bacterial and viral parasites at bay. The underlying principle is the following: as solitary multiplying forms, prokaryotes and viruses are forced by selection on fitness into a general growth contest. Thus they show fast exponential growth, whenever the situation allows it. For the generation sequences of the respective parasites this means that they require a relatively low probability of multiplication, in order to keep the overall growth at a level whereby their hosts will not become extinct. In other words, for sustainability reasons they need a niche situation in which favourable conditions can change very fast and frequently for the individual generation sequence. As a result, the generation sequences of the parasites need to find sufficiently "unstable" conditions, corresponding to their low level of complexity. Obviously their eukaryotic hosts are in this respect decisive, or more precisely, the capability of their immune defence to cause such an unstable situation for viral or bacterial invaders.

In more general terms, the above simply means that the eukaryotic immune defence varies enough between individual hosts, so that the parasites are sufficiently often confronted with conditions under which they cannot multiply. And exactly in this regard, the capability of eukaryotes would be decisive to conserve genetic information without a direct fitness benefit much longer. This corresponds to the basis for what can be described as a preventive defence against a multitude of potential parasite variants. This is documented by the fact that the eukaryotic immune system provides a vast diversity of antigen recognition (for the immune system see e.g. Campbell N.A. et al. 2008). At the end this corresponds to the preservation of genetic information without a fitness benefit over multiple generations, because the large majority of antigen recognition is not used by the individual form during its lifetime - I will come back in more detail to this point. As a consequence, this specific eukaryotic quality enables viral and bacterial parasites to sustain their generation sequences, without wiping out a much more complex and slower multiplying eukaryotic hosts.

Obviously the situation is different in the case of two non-eukaryotic forms, such as bacteriophages and prokaryotes. In this host-parasite relation, the prokaryotic host is considerably more complex than its respective bacteriophage (for phages see e.g. Clokie M. R.J. et al. 2011). Again, this relation can only then be sustainable, if the viral form does not wipe out its host. From the above argumentation it follows that the difference in complexity, and the related multiplication rate, cannot be too big. And indeed, in accordance with this demand, virulent bacteriophages, such as T4, which kill their host by lysis after infection, belong to the structurally by far most complex viral forms (Yap M.L. and Rossmann M.G. 2014). Obviously this fact lowers the difference in the structural complexity between the viral parasite and its prokaryotic host. It therefore corresponds to a support for the above conclusion that a negative coupling of the multiplication between solitary multiplying forms must face restrictions.

Last but not least, also the case must be taken into account in which both the host and the parasite are eukaryotes. With regard to the previous argument this situation corresponds to a struggle between "equals", in which the immune system of the host may not be able to keep the eukaryotic parasites at bay in the same way, as is the case with viruses and prokaryotes. Now, if the solution cannot come from the host, it must come from the parasite itself. Hence, the generation sequences of eukaryotic parasites need a self-made solution that avoids the extinction of their more complex hosts. The most obvious possibility in this respect would be an additional selection, not necessary for prokaryotic and viral forms. The general option for eukaryotes to cause an additional selection is via sexual reproduction and the mutual selection associated with it. This however is difficult for parasites, which are normally captured inside hosts and not able to compete with each other for mating or interaction partners (for sexual reproduction and parasites see e.g. Wesson R. 1991). Therefore they need another, particular solution. A conceivable way is to increase the risk of negative events during the individual life-cycle. Here the transition from one host to another corresponds to a critical point with multiple risks for the individual generation sequence of getting stopped. This means that the parasites would have to extend their life-cycle over different types of hosts. As a result, the likelihood for the individual generation sequence to get stopped would significantly increase. A corresponding life-cycle with more than one host would therefore lower the probability of multiplication of the parasite's generation sequences, in order to allow a sustainable host parasite relation. In other words, a lower probability of multiplication, which results from the transition between different hosts, can reduce the risk that the more complex hosts become extinct. And indeed, this is exactly what is found in nature, where eukaryotic parasites are often characterized by very complicated life-cycles with different host organisms (see again e.g. Wesson R. 1991).

In summary it can therefore be said that the observable characteristics of host-parasite relations do indeed support the conclusions drawn about complexity and biological niche factors.

1.9. A rising probability of multiplication

In the previous section I have discussed the relation between niche requirements and complexity. The resulting insight is that complexity increase needs the adaptation to irreversibly exhaustible niches with a relatively higher stability or reliability for the individual generation sequence. In this way the generation sequences of more complex forms can have a relatively higher probability of multiplication. This leads to the following conclusions.

Irrespective of the precise nature of the niche conditions, it can be assumed that the probability of multiplication of forms is dictated by certain aspects of their lifecycle. As an example, during a lifecycle cell-differentiated multi-cellular organisms go through a sensitive youth phase, before they reach the much more robust adult status. And the youth phase of animals and plants is obviously much more prone to negative generation sequences stopping events. Therefore it can be said that the early phases in the lifecycle of multi-cellular organisms correspond to a decisive weak point for the probability of multiplication. The chance that an individual generation sequence becomes stopped by whatever kind of event is much higher during this life stage than during the adult, fully developed phase – young rabbits are in greater danger of becoming the prey of foxes. Due to the relation between complexity and probability of multiplication the following can thus be deduced: complexity increase during evolution should occur along weak points of the probability of multiplication. This means that complexity increase would require the acquisition of qualities which strengthen these weak points. This conclusion has interesting consequences. If it is possible to identify a decisive weak point of the probability of multiplication for a particular type of form, it should allow general predictions about the next step of evolution towards increasing complexity. And indeed this is the case, as I will show in the following. The fact that such predictions are possible corresponds to a decisive quality of *Interaction Theory* compared to traditional evolutionary theory, which does not allow comparable predictions (for the latter see Lloyd E.A. 1994). For illustration, I will show in the following two examples, how weak points of the probability of multiplication can become strengthened.

The probability of multiplication at the beginning of evolution

The first example of how a weak point of the probability of multiplication may dictate the direction of complexity increase is for multiplying forms of very low complexity, as might have been the case at the beginning of evolution. In the next section I will then give an example for higher organisms.

In a supposed pre-biotic situation, in which relatively simple molecular forms become multiplied, the multiplication depends on the constant supply of factors such as suitable monomers, energy riche substances etc. (see e.g. De Duve C. 1995 or Kauffman S. 1995b). The mix of substances which needs to be supplied by the environment corresponds to essential resources necessary for the generation sequences of this type. For simplicity let's imagine as an example that the building blocks would correspond to the decisive resource for the multiplication of simple molecular forms. If the multiplying molecule would be proteins, the essential building blocks correspond to amino acids, such as alanine or tryptophan. Here it makes sense to assume that these building blocks are not available everywhere in the environment, and that the availability of some of them would in particular be challenging. This assumption fits well with the famous experiment by Miller (1953), who tried to imitate the early pre-biotic conditions on earth. With his trial setting, Miller produced a mix of molecules which could potentially be relevant at the beginning of life, such as several amino acids. And not surprisingly, not all of these molecules were produced with the same reliability (see Miller S.L. 1953). For the multiplication of a hypothetical protein that consists only of the amino acids alanine and tryptophan this means that the likelihood of the supply of a building block becoming disrupted depends on the availability of these amino acids. Here it makes sense to assume that the availability should in general be higher for a relatively simple molecule, such as alanine, compared to a more complicated molecule, such as tryptophan (for amino acids see e.g. Campbell N.A. et al. 2008). For the respective generation sequences this difference has consequences, because if one of the amino acids is no longer available, the generation sequences concerned will stop. This means that the substance whose availability is the least reliable will burden most on the probability of multiplication of the generation sequences. For the example of the hypothetical protein this has the consequence that if its generation sequences depend on alanine only, then they can own a relative higher probability of multiplication than with a dependence on alanine and tryptophan. Consequently, the dependence on the latter would correspond to a weak point for the probability of multiplication of this multiplying form.

This example can be transferred to the complexity increase during evolution as follows. If the evolution started with relatively simple, multiplying molecules, the

dependence of their multiplication on certain conditions and substances corresponded very probably to a general, decisive weak point for the probability of
multiplication. In accordance with the above argumentation, those conditions and
substances which had the highest risk of being suddenly no longer available
should therefore decide about the further direction of complexity increase. This
means that the evolution to more complex forms requires a strengthening of this
weak point, e.g. by acquiring the capability to multiply with substances which are
available with a relatively higher reliability. In other words, the insight that complexity increase goes in the direction of a rising probability of multiplication allows the conclusion that the early evolution has to follow the way from niches
characterized by a very high likelihood that the corresponding conditions became
disrupted, towards niches with a much lower respective likelihood. For the multiplying molecules at the beginning of evolution this would mean that they need to
evolve into forms which are, for instance, capable of synthesizing the essential
substances for their multiplication from relatively simple and reliably available
precursors. In this way, the probability of multiplication could increase for the
early generation sequences. The obvious consequence of this would be that the
cellular metabolism can be seen as the outcome of a corresponding development.
This in the sense that the basic metabolic pathways would have evolved step by
step, in order to ensure more and more reliable conditions for the multiplication
(see e.g. De Duve C. 1995 or Maynard Smith J. and Szathmáry E. 1995).

While looking at the cellular metabolism of living forms, it is striking that it is
organized around the interacting molecules DNA, RNA and protein. These molecules and their reproductive interaction can therefore be seen as the documentation of the common molecular history of all living forms[5]. Combining this with
what was said does mean that the evolution of the basic metabolism would have
been driven by the need to make the multiplication of these interacting molecules
more reliable – later I will come back to this point with regard to the definition of
life.

On the other hand, as a consequence of what was said, the direction of complexity
increase would be independent of the precise nature of the molecules that started
the evolution. Whatever those looked like, it can be assumed that their multiplication would have been dependent on very particular environmental conditions
(see e.g. De Duve C. 1995 or Kauffman S. 1995a). And a multiplication that depends on the presence of a very specific cocktail of substances in the surrounding
medium represents a weak point for the probability of multiplication of the respective generation sequences. Irrespective of the precise nature of the molecular

---

[5] For simplicity I'm not mentioning here the lipid bilayer, which can be seen as an additional
"interacting" entity dating back to the beginning of life – see later.

forms, the following can therefore be concluded about the complexity increase at the beginning of life: the starting evolution would go toward forms which have strengthened the decisive weak point of the probability of multiplication so that the average likelihood to continue over time would be higher for the individual generation sequence. The way how this is possible would be via progressively increasing biochemical capabilities allowing the use of continuously more reliable resources. Prokaryotes are the living forms characterized by outstanding biochemical capabilities. They are able to grow on nearly nothing and almost no organic substance exists that prokaryotes are not able to use as source of carbon and energy (Margulis L. 1993). For this reason, they can be seen as the direct and inevitable outcome of an early evolution that is directed towards an increasing probability of multiplication. They represent the ultimate endpoint of a development that is driven by a continuing strengthening of the general weak point of the probability of multiplication at the beginning of all generation sequences.

In summary, the insight that biological complexity increase is directed towards a rising probability of multiplication allows the following prediction about the beginning of evolution: it should correspond to a development that lowers the likelihood for the early generation sequences to become stopped, because of a low reliability of the required conditions for the multiplication. Biological forms able to grow on nearly nothing thanks to strong biochemical capabilities, such as prokaryotes, can therefore be seen as the inevitable outcome of this development. In other words, whenever and wherever multiplying molecules start an evolutionary auto-complexification, they should eventually produce forms comparable to prokaryotes, which can grow on very reliable and easily available resources – in the second part I will come back to this point.

The way from spawn to placenta

The second example of how the probability of multiplication can increase during evolution relates to higher organisms, such as species with sexual reproduction.

The lifecycle of sexually reproducing animals and plants is marked by different phases, such as zygote, embryo, youth and adult phase. These different parts of the lifecycle show important differences with regard to their environmental competence. For example, animals as well as plants are characterized by a touchy youth phase compared to a more robust adult situation. Thus, the non-adult parts of the lifecycle correspond to an obvious weak point for the probability of multiplication. The chance that a negative event will stop the individual generation sequence is significantly higher during this period in life than for the adult, fully developed organism. Consequently, the corresponding complexity increase should have to do with a strengthening of this weak point. As a consequence, more complex animals and plants should be characterized by qualities which decrease the likelihood for the individual generation sequence to become stopped during the non-adult phase. In summary, if complexity increase follows a rising probability of multiplication, the complexity increase of higher organisms should be characterized by the acquisition of qualities which provide relatively more stable or reliable conditions for the early, touchy phases of the lifecycle.

The evolution of species shows in fact this trend, as it is clearly visible, for instance, in the evolution of terrestrial vertebrates. Compared to the oceans, terrestrial life can be seen as more challenging by unpredictable climate changes, with negative implications in particular on the non-adult phase. From this point of view the course of evolution from amphibians with spawn, via reptiles with eggs, to mammals and placentalia is not surprising. In this progression the forms have acquired step by step qualities that can be seen as an investment into relatively more stable or reliable conditions for the pre-adult phase. In the same sense, the evolution of breeding birds also fits into this scheme. The egg protects the embryo better than spawn and provides more reliable growing conditions for the embryo. The brooding of the eggs provides more stable temperature conditions for the development and makes it more independent of climatic changes. The mammary gland ensures a reliable, optimal nutritional supply after birth and the placenta ensures the same for the embryo, as well as improved protection and mobility. In all cases the respective quality can reduce the likelihood for the individual generation sequence to be stopped in the early phase of the lifecycle. Similar arguments can be made about the evolution in other phylogenetic lines. So the evolution of terrestrial plants resulted in angiosperms with flowers and fruits. The latter are once again qualities connected with the touchy early stages in the lifecycle. Also the evolution of eusocial insects, such as ants and bees, fits into this scheme – later I will come back to these examples in more detail.

In summary, the insight that biological complexity increase follows a rising probability of multiplication allows predictions about the course of evolution. To become more complex, the forms need to strengthen a decisive weak point of the probability of multiplication in their lifecycle, in order to lower the risk of the individual generation sequence being stopped. In the case of higher organisms this means the acquisition of qualities providing relatively more stable or reliable conditions for the early phases in life. This conclusion is supported by the fact that the complexity increase of species can indeed be linked to the acquisition of corresponding qualities.

At this point it is necessary to highlight an obvious accord with the established, fitness-centred understanding of evolution. It is without doubt that selection on fitness can act on the weak points of the lifecycle. If young animals are an easy target for predators, then the acquisition of qualities which increase their fitness, by being better protected, provides a strong benefit for the multiplication success. Therefore it is not surprising that respective adaptations can be found in many species, such as the capability of young zebras or gnus to run away soon after their birth. Obviously corresponding fitness-enhancing qualities can also raise the likelihood for the individual generation sequence to continue. And this again may increase the probability of multiplication of a form. In this regard however, the following is important. As already indicated, a rising probability of multiplication is necessary, but not sufficient for complexity increase. In the next chapter, I will therefore clarify the difference between a purely fitness-driven increase in the probability of multiplication and one that results in a complexity increase. Before however, it is necessary to describe first the consequences of a rising probability of multiplication for the multiplication potential, which I will do in the rest of this section.

From quantity to quality

Interaction Theory considers the impact of increasing fitness on the environment. This in the sense that generation sequences in irreversibly exhaustible niches are not only endangered by negative environmental changes, they are also at risk of destroying the niche conditions by unsustainable growth. Therefore, if the decisive weak point in a lifecycle becomes increasingly robust, it can endanger a hitherto given sustainable balance with the environment. In other words, if fewer generation sequences are stopped per time, their number can rise up to a critical point at which the conditions are irreparably exhausted or destroyed. A strengthening of the weak points in the lifecycle in irreversibly exhaustible niches bears therefore the risk of resulting in an unsustainable number of generation sequences. Once more, this shows the fundamental dilemma that is associated with an evolution solely driven by selection on fitness: In a world with irreversibly exhaustible resources, a progressively increasing fitness that causes self-enhancing growth with accelerated resource consumption represents a fundamental risk.

On the other hand, the answer to how the probability of multiplication can rise in a sustainable way was already given: biological complexity increase has to go together with a decreasing multiplication potential. As a consequence, it can be concluded that for sustainability reasons, complexity increase demands not only a strengthening of a weak point in the lifecycle, but also a decreasing frequency with which the corresponding generation sequences ramify. In other words, complexity increase needs a development form quantity to quality, resulting in a higher investment into the single generation sequence. Hence it is possible to re-formulate the key insight of *Interaction Theory* as follows:

- The complexity increase of multiplying forms during evolution goes in the direction of a rising probability of multiplication of the concerned genera-tion sequences by a development from quantity to quality.

The application of this central insight to the evolution of species would mean the following. The investment into the early phases of the lifecycle, as described above, should not only cause a higher likelihood for the individual generation se-quence to proceed over time. To be sustainable, it should also follow a develop-ment from quantity to quality, what is indeed the case - see below.

Decreasing branching attempts of generation sequences

That complexity increase follows a development from quantity to quality can eas-ily be seen with unicellular organisms. Here, each successful cell-division corre-sponds to a ramification of the respective generation sequence. A development from quantity to quality means therefore fewer cell divisions of the individual form per time. And indeed, more complex unicellular forms, such as protozoa, need significantly longer for a cell cycle than prokaryotes (see e.g. Campbell N.A. et al. 2008). Thus, the generation sequences of more complex eukaryotic forms show less branching attempts per time under favourable conditions, compared to structurally less complex prokaryotes. It can therefore be said that the complexity increase of unicellular organisms follows indeed a development from quantity to quality, measured by the time between two consecutive cell divisions under opti-mum growth conditions. For this type of organism, the corresponding time differ-ence can therefore be seen as a proxy for the difference in the probability of mul-tiplication and thus the complexity.

With the emergence of sexually reproducing multi-cellular organisms, the branch-ing attempts of generation sequences depend on more than a cell division. So, different taxonomic clusters have different kinds of reproduction, in the sense that for example amphibians spawn and reptiles lay eggs. On the other hand, in all cases the multiplication potential of sexually reproducing animals or plants de-pends on the number of fertilized egg-cells they can produce per time. Each ferti-lized egg or zygote can be seen as a branching attempt of the respective generation sequence that, if successful, will contribute to an increasing number of generation sequences. Consequently, for a sustainably rising probability of multiplication, by a development from quantity to quality, the number of zygotes that are produced per individual per time under favourable conditions should decrease.

To verify this conclusion, let's first look at the macro-evolution of multi-cellular organisms and ask whether the number of branching attempts per time decreases over the course of evolution? Here it is necessary to consider that for multi-cellular organisms the multiplication potential is influenced not only by the mode of reproduction, but also by body size. For example, a mouse can multiply much faster than an elephant, because the smaller size allows a much faster development. On the other hand, a big amphibian can produce a high amount of spawn and thus more zygotes per time than a little bird by laying eggs. As a consequence, the number of branching attempts per time can only serve as an indicator for the relative complexity of sexually reproducing animals and plants, if the impact of body size and reproduction mode is considered. This can be done in two ways. Either by comparing organisms of similar size, such as a small frog with a small lizard; or the average branching attempts can be compared between different taxonomic groups with different modes of reproduction, such as between reptiles and mammals. The first allows a comparison of individual organisms, while the second with regard to the so called macro-evolution.

With this in mind let's ask what has happened to the number of multiplication attempts during the evolution of species? Here I must admit that my knowledge of zoology and botany is rather limited. Nevertheless it seems obvious to me that more complex forms show in general fewer branching attempts per time, and I trust that most readers will share my view. For example, if evolution is considered at large, so there is little doubt that relatively simple organisms, such as worms, multiply much faster and show therefore significantly more branching attempts per time than later emerging, more complex organisms, such as reptiles, to keep a sustainable balance with their environment. The same is true if the branching attempts of unicellular organisms are compared to those of cell differentiated multi-cellular organisms. Therefore it is fair to say that evolution has followed by and large the expected development from quantity to quality.

The same is however true, if more closely related groups are compared. Let's take for example the evolution of terrestrial vertebrates. Again, most readers will very probably agree that amphibians generally make significantly more branching attempts per time by spawning, than reptiles by laying eggs, and that the latter for their part lay in general more eggs per time than mammals produce embryos. With regard to mammals it is also striking that some forms evolved towards an increasingly longer generation time. The inevitable consequence for the generation sequences of these species is that the time between consecutive ramifications becomes significantly longer. With the modern humans and a generation time of more than 20 years, this development has reached an unprecedented level. It is therefore possible to say that the phylogenetic line of terrestrial vertebrates to

Homo sapiens shows by and large an overall decreasing number of branching attempts per time and therefore a development from quantity to quality. And in my opinion the same can be said about other branches of the phylogenetic tree. For instance, I have no doubt that birds produce in general fewer eggs than their phylogenetic predecessors. Other examples would be the comparison of algae with moss or gymnosperms with angiosperms or primitive insects with eusocial insects. In all cases, the phylogenetically more advanced group shows on average less branching attempts per time. As a consequence, the conclusion that biological complexity increase has to follow a development from quantity to quality is true for so-called macro-evolution, at least according to my personal knowledge (for a great overview about the biodiversity and the phylogenetic tree of life see "The Tree of Life web project").

To avoid any misunderstanding, the overall trend does not mean that it is true in every single case. For example, even if mammals show in general less branching attempts per time than reptiles, it does not mean that this is true for each mammal species compared to any type of reptile. And this makes sense if the fact is considered that species are adapted to different niche situations, which might demand different multiplication rate. This implies also that it would not make sense to directly compare very different kind of organisms, such as mammals and angiosperms, with completely different ways of living. This again means that if the multiplication attempts of groups are directly compared, they should be in the same phylogenetic lineage, as is the case for terrestrial vertebrates.

The above raises the question of whether, the relative number of branching attempts can be used as measure for the relative complexity? As said, first of all it would be necessary to compare organisms with similar size. For example, regarding the average numbers of branching attempts per generation, an elephant can be compared with a dinosaur of comparable size. And if the number of branching attempts is decreasing in the course of evolution, the dinosaur would on average have laid more eggs per time than elephants give birth. I am confident that this is true, even if I couldn't find a reference for the average number of dinosaur eggs in literature. The fact that elephants have only one offspring at a time, and the female is not receptive during the following years while they nurse their young, can be seen as a further argument. In addition, young elephants may need longer to reach sexual maturity than dinosaurs of comparable size (again I couldn't find a reference in the literature). Hence, it is very likely that elephants show less branching attempts per time than a dinosaur of comparable size. If the number of attempts serves as a measure for the probability of multiplication, and thus for the relative complexity, this mean that from the *Interaction Theory* point of view, the complexity of an elephant is relatively higher than that of dinosaurs of similar size. If this is extrapolated it would mean that mammals as a genus can be seen on average as more complex than the genus dinosaurs.

A similar comparison can be made between angiosperms and gymnosperms, for example regarding an apple tree and a spruce of comparable size. In this case, the difference in branching attempts per generation is obvious. The apple tree reproduces via sophisticated blossoms and apples, while the spruce with comparatively simple seeds. If all apple grains produced during a season, were counted and compared to the average number of seeds produces by the spruce, the difference should be enormous. As a result, from the *Interaction Theory* point of view this would mean that apple trees are relatively more complex than spruces. And if this is again extrapolated it would mean that angiosperms are on average more complex than gymnosperms.

In summary, the comparison along the phylogenetic tree of life supports the conclusion that complexity increase is accompanied by a reduction of the branching attempts of the respective generation sequences. This is in line with the central dogma of *Interaction Theory*, namely that complexity increase during evolution goes into the direction of a rising probability of multiplication, in the sense of a development from quantity to quality.

1.10. The way in which the probability of multiplication can rise

After having clarified the basic principles as to how the probability of multiplication can rise in a sustainable way, I will now discuss the process by which this can occur during evolution. Here it turns out that two possible ways exist, both of which have different implications for evolution. To come to this conclusion, I need first to mention two points that are central in this regard.

Benefit gradients and phylogenetic clusters

The development and fixation of new qualities or traits during evolution requires an adequate selection (Maynard Smith J. and Szathmáry E. 1995). This should also be the case for the qualities necessary for a rising probability of multiplication. This means that variants with respective, inheritable qualities or traits prevail over time, because of a relatively higher multiplication success than their peers. In the two examples of the previous chapters, the rising probability of multiplication resulted from the capability to synthesize amino acids, such as tryptophan, from more reliably available substances, or from new anatomic structures, such as the mammary gland or the placenta. The first needs a set of new enzymatic activities, the second multiple metabolic and anatomic changes. In both cases, the acquisition of the new qualities can therefore not be a trivial, straightforward task, but requires several steps over multiple generations. Thus, it makes sense to assume that the emergence and fixation of qualities which cause a rising probability of multiplication can in general not occur in a single variation or mutation step. Instead, corresponding qualities must develop in a consecutive succession of steps, so that they eventually emerge. Consequently, a rising probability of multiplication via a development from quantity to quality would be the result of a selection on inheritable variants over multiple generations. During this process, the respective variants must own a tangible and sufficient advantage compared to their peers, so that they can prevail. This can be expressed by the following demand: the development and fixation of qualities for a rising probability of multiplication requires a suitable benefit gradient over multiple generations.

The second point I want to address is the organization of the phylogenetic tree of life and the fact that living forms do not show a gradual but a clustered biodiversity (e.g. Stearns S.C. and Hoekstra R.F. 2005). The difference between clusters or groups, such as reptiles and mammals, is based on the possession of distinctive traits. In other words, all mammals have specific qualities in common which reptiles do not own, and vice versa. The same can be said regarding gymnosperms and angiosperms, or prokaryotes and eukaryotes etc.. In comparison, the diversity within a cluster corresponds to the variation of a common theme. This means that the members of a particular cluster can be distinguished by species specific features, such as colour and shape of the plumage in the case of birds. But regardless of these phenotypic differences, the species within a phylogenetic group have particular traits in common, which make them recognizable as members of this specific cluster (see e.g. Gee H. 2000).

Closely linked to the clustered organisation of the biological diversity is the fact that also the complexity increase of living forms did not occur in a gradual manner within a particular phylogenetic group. The evolutionary auto-complexification has led rather to the emergence of ever new taxonomic groups with increasingly complex forms (see e.g. Conway Morris S. 2003). In other words, within a phylogenetic line the newest taxonomic group shows the on average more complex forms (see later). This leads to another important conclusion:

- Complexity increase during evolution results in a phylogenetic tree of life characterized by a clustered biological diversity, with the most recently evolving clusters harbouring the on average more complex forms.

The explanation sought as to why and how the biological complexity increased during evolution must therefore not only be in accordance with this fact, but also provide a satisfactory explanation for the clustered biodiversity of life as such.

For completeness, it should be mentioned that the relevant explanations provided by the conventional theory of evolution are not satisfactory (see e.g. Eldredge N. 1995).

Selection on a rising probability of multiplication

Since complexity increase during evolution would demand a rising probability of multiplication by a development from quantity to quality, let's now discuss the principles by which this would be possible.

The main insights regarding complexity increase are so far that the individual generation sequence of more complex and therefore slower multiplying forms needs relatively more stable or reliable niche conditions. This becomes visible in a relatively higher probability of multiplication of the corresponding generation sequences. For sustainability reasons, a greater likelihood of the individual generation sequence to continue over time must go along with a development from quantity to quality. This means fewer ramifications of the generation sequences along with a higher investment into the single progeny. The way in which this is possible is via a strengthening of the weak point for the probability of multiplication in the lifecycle of the respective forms. In other words, to be more complex, forms need to acquire qualities that strengthen the phase in the respective lifecycle where most generation sequences are stopped.

Based on these insights, let's now clarify the kind of selection that may cause the development of corresponding qualities. Here, a difference between generation sequence and multiplying form becomes obvious. On the one hand, for any kind of selection the multiplication success of the individual multiplying form – be it a molecule, cell or organism – is decisive. On the other hand, the probability of multiplication was introduced in connection with generation sequences. It expresses the average likelihood of a multiplying form to be part of a generation sequence, which can continue and will not be stopped. A correspondingly increasing likelihood as such, however, does not automatically mean a direct benefit for the individual form and its multiplication success in the here and now. This is comparable to some kind of investment which provides only some generations later, i.e. in the distant future, with a benefit. In other words, if something is good for the generation sequence, it does not mean it is also good for the single form (for respective arguments see again e.g. Williams G.C. 1966). For example, why should a frog invest more into the single tadpole and reduce in parallel their total number, corresponding to a development from quantity to quality? For the frog the multiplication success relative to its peers is decisive, and as long as it is sufficiently high with the relatively cheap production of tadpoles, there is no immediate benefit in changing this strategy. On the contrary, the production of a high number of tadpoles has the advantage that the individual frog can maximise its multiplication when many tadpoles survive because of very favourable conditions. As a consequence it can be assumed that a development from quantity to quality requires particular circumstances.

So then, how can a development from quantity to quality happen? The answer is that it can happen in two different ways, either via a mono or a dual mode of selection. Mono means in this context that only selection on fitness is needed for the development from quantity to quality, because a corresponding benefit gradient is given. In other words, in the succession of generations from the starting to the endpoint of the development, each of the corresponding intermediate variants requires a relatively higher fitness and therefore a higher multiplication success than their peers. In comparison, dual stands for the consecutive action of mutual selection and selection on fitness. Only together can both types of selection provide the necessary benefit gradient. More precisely, the dual mode starts with mutual selection that provides a first, non-fitness related benefit gradient. At some point, selection on fitness comes into play and provides a subsequent, second benefit gradient, by which the mutually selected qualities become transformed into the final form. As I will explain in the following, it is this dual mode of selection which would be responsible for complexity increase during evolution. First however, let's discuss how selection on fitness alone can cause a development from quantity to quality and thus a rising probability of multiplication.

A good example for a selection on fitness driven development from quantity to quality is that of an increasing body size. As already mentioned previously, big animals such as elephants cannot reproduce as fast as small animals, such as rats or mice. On the other hand, with their much bigger size the individual generation sequence of elephants is also less in danger of getting stopped, because the fragile youth phase is generally better protected. This means that if a species becomes bigger during evolution, as a result of selection on fitness, it would raise the likelihood for the individual generation sequence not to get stopped, and at the same time reduce the multiplication potential.

The second example concerns mouthbrooder fishes, which take their brood into the mouth (see e.g. Wesson R. 1991). Obviously this behaviour provides a protection for the fragile youth phase and should therefore result in a rising probability of multiplication. It is also obvious that the mouth can only harbour a limited number of offspring, which represents a limitation for the multiplication potential of these fishes. As a result, the evolution of mouthbrooder fish can also be seen as a development form quantity to quality. Relevant for the discussion is that with regard to both examples, increasing body size and mouthbrooder, it can be assumed that the development is the result of a gradual selection on fitness.

In comparison, good examples of a development from quantity to quality driven by a dual mode of selection would be the development of spawning amphibian into egg-laying reptiles, or the latter into mammals. Before detailing the reason why these examples would need a dual mode of selection, I would first like to highlight an easily visible key difference between the two ways of selection. In case the development is driven solely by selection on fitness, it leads normally not to a new phylogenetic cluster. So an increase in body size has occurred in many phylogenetic groups, and mouthbrooders evolved independently in different families of fish (see e.g. Helfman, G.et al. 1997). By contrast, a development from quantity to quality driven by a dual mode of selection would inevitably lead to a new phylogenetic cluster, distinguishable from the hitherto existing diversity – as for example mammals formed a new phylogenetic group that is different from their reptile-like ancestors. This has the important consequence that the dual mode of selection can provide an explanation of why the phylogenetic tree of life has a clustered organisation – see below.

The adaptation to new types of niches

Why would the dual mode of selection result in a new phylogenetic cluster, while an exclusive selection on fitness does not? To answer this question it is necessary to consider that beside self-made causes, generation sequences can be stopped by two different kinds of negative events. Either the individual form can be hit, for instance, by a fatal lightning strike or a deadly encounter with a beast of prey. Or all respective forms are hit at the same time by an environmental change, causing for instance a climatic change with drought or unstable temperature conditions. To strengthen a weak point in the lifecycle against the first kind of event can in principle be a straightforward affair. For example, if variants with a bigger size have a relatively higher multiplication success, because they are better protected against predators, it can be expected that the respective form becomes bigger. How far such a development can go, and how big the forms eventually become, would depend on the particular circumstances. That is to say that the development can go up to a point from which a further increase in size is no longer an advantage. Similar arguments can be made for the mouthbrooder example. In both cases it would depend on the particular niche situation, if and how far a gradual selection on fitness can cause the respective development. In the end it means that an exclusive selection on fitness-driven development from quantity to quality can then occur when the multiplication success and the probability of multiplication are directly connected.

76

In contrast, such a direct connection between multiplication success and probability of multiplication would not exist in the case of a dual mode of selection. Let's take, for example, the transition from a reproduction with spawning to one with eggs. The strengthening of the weak point in the lifecycle, which allows a development from quantity to quality, comes in this case from a substantial change on the level of niche conditions. This means that independent of the particular niche, the situation for reptiles differs in general at least with regard to one niche factor which is essential for amphibians. The latter need something like a water pond for their reproduction via tadpoles, a condition that is no longer necessary for egg-laying forms. The same argument can be made for other respective developments, such as the development of forms with regulated temperature metabolism from cold-blooded ancestors. In this case the dependence on climatic conditions has fundamentally changed, because thermo-regulated mammals and birds can reproduce in a much wider temperature range than their cold-blooded ancestors.

Decisive with regard to the further discussion is the assumption that this kind of change on the level of niche factors or conditions cannot in principle be the result of an exclusive selection on fitness. The reason is the following: for niche factors or conditions that are essential for the multiplication, selection on fitness might cause a better adaptation, as long it goes along with a relatively higher multiplication success. In this way a frog may, for example, improve the survival rate of its tadpoles. It is however, something completely different to demand a development over multiple generations so that the multiplication becomes eventually independent of an essential niche factor or conditions. In the example this means that selection on fitness would need to transform a frog into a reptile, so that the multiplication is no longer dependent on water for spawning. The existence of a corresponding gradual benefit gradient for selection on fitness appears hardly plausible, because it does not demand an improvement of something existing, but rather an investment into something new. If it were otherwise, intermediate forms between amphibian and reptile should also be expected (see e.g. Eldredge N. 1995). This insight leads to a further demand of *Interaction Theory*, which reads as follows:

- An exclusive selection on fitness cannot make forms independent of an essential niche factor or condition, if the independence is not possible in an accidental mutation step, but rather requires a successive development over multiple generations.

Hence, an amphibian could not be transformed into a reptile only by adapting to the environment, or a gymnosperm could also not transform in this way into an angiosperm etc. To become independent of an essential niche factor or condition would rather require that the form invests first over multiple generations into a new quality, which is, however, only beneficial when the process is widely completed. And a respective investment should not be possible by selection on fitness alone. But how else can such a development happen? The answer already given is that it would require the combined action of mutual selection and selection on fitness, in the sense of a dual mode of selection. The transformation of an amphibian into a reptile or a gymnosperm into an angiosperm would therefore demand the successive action of mutual selection and selection on fitness.

As mentioned at the end of the last section, an important consequence of this role of dual selection is that the clustered organisation of biological diversity becomes explainable. The explanation is the following: if as a result of dual selection a form is suddenly able to multiply under fundamentally different environmental conditions, then it can be expected that this form will also become the origin of a new type of diversity. Thus, a new taxonomic cluster emerges. In other words, the acquired capability to multiply under substantially different environmental conditions would allow the adaptation to a new kind of niche not accessible for the other existing forms. As an example, in a world governed by cold-blooded dinosaurs, the first mammals emerged as new taxonomic cluster, because their specific characteristics gave them access to a new kind of niche. More specifically, the thermo-regulation would have allowed the first mammals to be nocturnal, which was not possible for dinosaurs of comparable size, at least not in the same manner (see e.g. Conway Morris S. 2003). In a similar way, the first reptiles were able to conquer dry habitats, not accessible to amphibians or the first angiosperms occupied habitats not suitable for gymnosperms, etc. (Soltis D.E. et al. 2008).

In summary, from the *Interaction Theory* point of view, complexity increase needs a development from quantity to quality by strengthening a weak point in the life cycle. This is possible by acquiring qualities that allow an independence of so far essential factors or conditions. For reasons of principle however, this cannot happen by selection on fitness alone. It would rather require a dual mode of selection, consisting of mutual selection followed by selection on fitness. This would be visible in the emergence of new taxonomic groups which are characterized by a capability of multiplying under substantially new conditions and can therefore occupy a new kind of niche. This conclusion is in line with the clustered organization of the phylogenetic tree of life.

The importance of key innovations for biological complexity

Now, how exactly does mutual selection function? In this regard, I will introduce the term key innovation. It describes newly emerging qualities during evolution, which cause a rising probability of multiplication as a result of mutual selection.

Key innovations are particular, because they would be responsible for the clustered biodiversity in the phylogenetic tree. Key innovations are therefore the qualities that make a eukaryote a eukaryote, a reptile a reptile or an angiosperm an angiosperm etc. From the *Interaction Theory* point of view, all these qualities have in common that they are different from ordinary environmental adaptations, because their acquisition would not be possible either by an accidental mutation steps, or by an exclusive selection on fitness. To emerge during evolution, they need to develop over multiple generations, as a result of a dual selection.

The introduction of key innovations can now be combined with what was previously said about the way in which the probability of multiplication can rise, and leads to the following demand:

- Key innovations make the multiplication independent of a formerly essential niche factor or condition with limited reliability, and so strengthen a weak point in the lifecycle where many generation sequences get stopped.

For this reason key innovations are closely linked to the process of complexity increase. The acquisition of a new key innovation during evolution would result in the appearance of a new taxonomic cluster in the phylogenetic tree of life, such as the first mammals appeared in a world dominated by dinosaurs or angiosperms in a world of gymnosperms. The new taxonomic group would be characterized by their independence from a formerly essential niche factor or condition. To stick with the example of angiosperms, from the *Interaction Theory* point of view these forms are no longer dependent on the factors or conditions necessary for the proliferation of the pollen and seeds of gymnosperms. Thanks to the key innovation of flowers and fruits, the individual generation sequence of angiosperms would have, on average, a relatively higher likelihood to continue[6]. Therefore their generation sequences would need relatively less branching attempts than those of gymnosperms. Other examples were already mentioned previously, in the form of the key innovations that differentiate reptiles or mammals from their respective, phylogenetic ancestor.

From the evolutionary point of view, the key innovations, as defined above, are something new and different compared to what existed before. The structure of the phylogenetic tree suggests that the acquisition of this kind of innovation happens relatively seldom (see e.g. Eldredge N. 1995 or Gould S. J. 2002). This

---

[6] Angiosperms need insects or animals for their proliferation. Hence, strictly speaking, angiosperms substitute certain factors or conditions by others. However, it is essential that the new factors or conditions are significantly more reliable than those required for the proliferation of pollen and seeds of gymnosperms.

would explain why the classification of living forms in taxonomic clusters is possible in the first place. The possession or non-possession of a respective key innovation makes an organism either a reptile or a mammal, while semi-reptiles or semi-mammals do not normally exist. In some cases, the steps may be smaller than in others and intermediate groups are possible, such as lungfishes as a link between fishes and terrestrial vertebrates. In other words, the so-called macroevolution is visibly marked by the emergence of key innovations, which are more than the ordinary environmental adaptations that differentiate the forms within a taxonomic cluster.

Central from the *Interaction Theory* point of view is that the emergence of ever new key innovations in the phylogenetic tree of life goes along with a continuously rising probability of multiplication in the course of the evolutionary complexity increase. This means that key innovations would have a positive effect on the weak point in the lifecycle and are therefore responsible for an increasing likelihood by which the individual generation sequence of more complex forms continues over time. Hence, it would be the acquisition of key innovations during the course of evolution that drives an increase in the probability of multiplication. The acquisition of key innovations can therefore be seen as the motor of complexity increase. In the end this means that complexity increase during evolution occurs by the stepwise accumulation of key innovations.

As a direct consequence of the central role of key innovations, it becomes possible to give a new meaning to biological complexity. It complements the prior definition from the beginning of the book, which was purely descriptive. This meaning is:

- The complexity of a biological form is the manifestation of the totality of its key innovations that have been acquired during its phylogenetic evolution.

In the following I will now discuss the decisive difference of key innovations compared to other adaptations. This is that they can only emerge as a result of a dual selection, but not via an accidental mutation step or an exclusive selection on fitness. As a reminder, dual selection means that in a first step it would require mutual selection for the development of a non-fitness-related precursor quality, which then becomes modified by selection on fitness into the final key innovation. In the following I will therefore also discuss how likely it is that corresponding key innovations emerge during evolution.

Key innovations as a result of mutual selection

The reason why key innovations would require dual selection is that not all of the physiological and morphological changes that are required for their development can emerge because they cause a relatively higher individual fitness. Hence, mutual selection has first to provide a non-fitness-related precursor quality. Since mutual selection cannot act purposefully, suitable precursors can only emerge accidentally. As a result key innovations would not develop easily, because this first step acts as a barrier. A way to overcome this hurdle would be that evolution produces constantly a high variety of non-fitness-related qualities. This would improve the chance that also a suitable precursor for a key innovation will emerge over time. As said, a suitable precursor means that selection on fitness can act upon the corresponding quality and modify it, so that it becomes a key innovation. It is this mandatory need for a non-fitness-related precursor quality which differentiates key innovations from other biological traits.

How can these precursor qualities emerge during evolution? Regarding their source or origin it would be necessary to differentiate between the two fundamental types of niches. In the case of niches at risk of becoming irreversibly exhausted or destroyed by growth, the sustainable multiplication would require mutual selection. This must result in the emergence of interaction related traits which can be seen as non-fitness-related qualities. At some point these qualities may become relevant for the individual fitness and would therefore correspond to potential precursors for key innovations. Given the mentioned importance of mutual selection for the sustainability of generation sequences, this could occur to varying degrees in all irreversibly exhaustible niches. The result would be a constantly available pool of mutually selected qualities. Over deep time, some very few of these qualities may eventually be transformed by selection on fitness into a key innovation.

Now, if these conclusions are correct, the constant availability of manifold non-fitness-related qualities must be visible in nature. And indeed this is the case and well visible with regard to sexual reproduction, the main form of reproductive interaction between higher organisms. In many animal groups the differences between species can be linked to characters directly or indirectly linked to sexual reproduction (Wesson R. 1993), thus qualities related to mutual selection. Within many non-vertebrates for example, genitals correspond to the most complex anatomic details. Most of these sophisticated qualities related to sexual interaction have little or no relevance for the adaptation to the respective niche (again Wesson R. 1993). In addition it is well known from livestock breeding that the selection on a new trait brings along other qualities, so-called free riders. With regard to the consequence of mutual selection this means that it would also result in such free riding traits and qualities, which appear together with the mutually selected quality. In the end, these kinds of findings support the assumption that mutual selection produces constantly a pool of non-fitness-related qualities, and thus potential precursors for key innovations.

As said, niches tolerating exponential growth do not require interaction communities with mutual selection. Here the situation is obviously different. In this case, the necessary precursor which selection on fitness can transform into a key innovation would originate from symbiosis – see below.

The consequences of interactions and symbioses

A main characteristic of the reproductive interactions introduced is that they cause a mutual selection, because in each generation the individuals with the relatively highest interaction competence have potentially an advantage with regard to their multiplication success. This corresponds to an ongoing selection on particular qualities that is present in each generation. As said, the precise outcome of this ongoing selection would be unpredictable and solely depend on the accidental variant composition of the particular interaction community and their specific preferences with regard to interaction competence. As a consequence, the ongoing mutual selection, e.g. in the form of sexual selection, can produce impressive qualities with no obvious fitness benefit, such as the striking plumage of many birds or the exotic form and colour of many flowers.

Compared to mutual selection, the niche specific selection on fitness would go into the direction of a better adaptation to given environmental conditions, i.e. towards variants with relatively more progenies per time under these conditions. In this regard, selection on fitness would be predictable, at least in principle - see the previous discussion with regard to the Spiegelman trial.

The postulation that the qualities caused by mutual selection are per se not relevant for fitness does not exclude that at some point they can start influencing the fitness, either positively or negatively (for the impact of sexual selection on fitness compare e.g. Maynard Smith J. and Szathmáry E. 1995). In other words, the ongoing selection caused by reproductive interactions can be seen as a constant source of new, unpredictable qualities, which can eventually impact the fitness of the respective form. In some cases this impact may be relatively strong, while in others it may be minor or negligible (for sexual selection see e.g. Miller G. 2001). As a result it can be expected that selection on fitness will join in and start modifying the respective quality, in particular if the fitness impact is strong. This allows the conclusion that in cases with a minor impact, the interaction-related origin of a quality may still be evident, while in cases where the impact on the fitness is strong, it may no longer be obvious. Here, selection on fitness may cover the interaction origin. The latter would in particular be the case for the above-mentioned free riders.

In view of what was previously said, it raises the question of how likely is it that qualities originating from a respective dual selection become key innovations? In this regard, it can be expected that in the overwhelming majority of cases this would not happen. Because even if a dually selected feature has a positive influence on the fitness of a form, it may normally not allow the multiplication to become independent of a formerly essential niche factor or condition. And because the latter was introduced as essential for key innovations, a corresponding dually selected quality would not be relevant for complexity increase. On the other hand, if a quality originating from dual selection has an impact on the environmental adaptation it might well cause the emergence of a new phylogenetic sub-cluster; this, when the impact on the fitness allows a diversification of the form that acquired the corresponding, dually selected quality. For example, the emergence of sub-clusters, such as the dog- or cat-like group, could be explained in this way. From the *Interaction Theory* point of view, the clustered organisation of the phylogenetic tree of life is therefore a visible sign of how important dual selection is for evolution. In most cases, this dual selection would cause the emergence of new phylogenetic sub-groups and contribute in this way to the strongly clustered organization of biological diversity. Yet, in some few cases, over deep time the dual selection may also result in a key innovation that strengthen a weak point in the lifecycle by making the multiplication independent of a formerly essential niche factor or condition with limited reliability. And it would be these few cases which drive the complexity increase during evolution.

In a similar way it can be assumed that symbiosis will cause the emergence of new qualities which would otherwise not emerge. Within unicellular organisms for example, the physical unification of symbiotic partners can produce new qualities not directly related to the original benefit associated with the symbiosis. The process would be as follows: different types of unicellular organisms come closely together as symbiosis partner, driven by the fitness related benefit of the symbiosis. If this results eventually in the physical unification of the hitherto independent forms in a new, larger entity, it can have consequences beyond the original symbiosis. A key example in this respect would be the increasing amount of DNA per cell, as has occurred during the complexity increase from prokaryotic to eukaryotic cells. As I will explain in more detail later, the significantly larger genomes of eukaryotes, compared to prokaryotes, can be seen as a quality that allows the multiplication under fundamentally different conditions. The way in which the first cells with a respectively increasing amount of DNA emerged, however, would have been the by-product of a physical unification of different prokaryotic cells engaged in a symbiotic relation. In other words, the new capability to preserve a higher DNA amount per cell would not have been the result of an exclusive selection on fitness, but rather the accidental consequence of the unification of formerly independent cells. Hence an essential quality of eukaryotic cells would have originated as an accidental by-product of symbiosis. The corresponding quality was then transformed by selection on fitness into the key innovation that is characteristic for eukaryotes (for the role of symbiosis in the emergence of eukaryotes see Margulis L. 1993).

In summary, the mutual selection on interaction competence, as well as that between symbiotic partners, can cause the emergence of new qualities which would not emerge by an exclusive selection on fitness. The consequence hereof is a permanently emerging pool of mutually selected qualities, out of which some few may eventually become key innovations. This means that they allow the respective form to become independent of a formerly essential niche factor or condition, so that it results in a relatively more stable or reliable situation for the corresponding individual generation sequence. In this way, reproductive interactions or symbiosis might drive the complexity increase of life. In addition, the impact of dual selection on the environmental adaptation in general can be seen as the reason for the characteristic clustered organization of the phylogenetic tree of life. This view is substantially different from the prevailing belief of a fitness-centred evolution.

The emergence of key innovations within established ecosystems

Next, let's ask what happens to generation sequences if non-fitness-related qualities become transformed by selection on fitness into key innovations?

If undisturbed by environmental changes, ecosystems would develop over time into what can be called established ecosystems. The term established expresses the situation that all available and accessible niches are occupied by the existing biodiversity. This niche occupation would be driven by selection on fitness and comes obviously to an end after suitable niches are no longer available. As a result, established ecosystems should show a relatively unchanging biodiversity, at least as long as the environmental conditions remain stable. This in turn should be visible in the fossil record by long periods of relative stasis that are punctuated by periods of rapid change, because of changing environmental conditions. Such a pattern corresponds to the so-called punctuated equilibrium which was proposed by Gould and Eldridge (1977). In the context of newly emerging key innovations this would mean that they should rather appear within established ecosystems during the long periods of relative stasis. The reason would be that the necessary, non-fitness-related precursor qualities would depend on mutual selection. And as previously discussed, this type of selection would in particular be important for the multiplication success during periods of relative environmental stability. In contrast, selection of fitness would become dominant during periods of changing environmental conditions.

What would that all mean for the emergence of key innovations? First of all it was said that key innovation-owning forms have an altered dependence on environmental factors or conditions that were so far essential for the multiplication. In other words, thanks to a key innovation it becomes possible to multiply under environmental conditions which are not possible for the respective direct ancestors. The obvious consequence thereof should be that at the moment when a key innovation emerges, the corresponding form would be able to occupy a new kind of niche in the respective ecosystem. This can also be expressed as follows: thanks to an acquired independence of a formerly essential niche factor or condition, the generation sequences of key innovation-owning forms can occupy a new, previously unoccupied kind of niche.

Furthermore it can be concluded that over time the key innovation-possessing forms become the origin of a new diversity within the established ecosystem, which is visible as a new taxonomic cluster. This is because over time selection on fitness would adapt the corresponding forms to those environmental situations under which their altered dependence on niche factors or conditions represents an advantage. However, this new diversity should remain limited. The reason is that the capacity to occupy a new kind of niche does not mean that the key innovation-owning forms can also oust other forms from their niches to which they are well adapted. The new diversity of key innovation-owning forms should therefore adapt to special situations and thus play a minor role in the established ecosystem. And indeed, this scenario corresponds to what can be found in the paleontological record. For example, the first mammals appeared during the reign of the dinosaurs and could only form a relatively limited diversity (see e.g. Eldredge N. 1995). Another example would be the minor role of the first angiosperms in a world dominated by gymnosperms (see e.g. Soltis D.E. et al. 2008).

In this regard, I would like to mention again an important point related to dual selection. If an originally mutually selected quality becomes adapted by selection on fitness so that it allows eventually the multiplication under new niche conditions, the origin might no longer be recognizable. This in the sense that the modifications necessary for the adaptation to the new niche situation may change the originally interaction- or symbiosis-related character of the key innovation so far that it is no longer obvious.

In summary, it can be assumed that key innovations emerge in established eco-systems where they allow the occupation of a new kind of niche. These niches are characterized by an independence of a formerly essential niche factor or condition. Therefore they cannot be occupied by similar forms, but without the corresponding key innovation. The resulting new diversity of key innovation-owning forms would however be limited to those environmental situations where the key innovation provides actually a tangible benefit. And as a result of the occupation of the new kind of niche, the mutually selected origin of the key innovation might no longer be recognizable.

Established ecosystems and the above scenario would be the result of an environment that is by and large stable. Over deep time however, this is obviously not always the case. In the following I will therefore take a closer look at the behaviour of the environment over long periods of time.

1.11. Environmental changes and complexity increase

In the discussion of the probability of multiplication and the meaning of niches, the factor time plays an essential role. For example, niches were introduced as islands of relative stability which allow the sustainable multiplication of forms in a given environment. The relative stability is seen as necessary, so that the respective generation sequences can proceed over a longer period of time. And longer means in this context a multiple of the generation time. When it comes to the evolution of biological complexity however, the factor time must be considered even in geological periods. So, the biological generation sequences on earth have been on their way through deep time since very probably almost 3.5 billion years (Knoll A.H. 2003). This is qualitatively different from the question of whether certain conditions can serve as a niche or not. For example, a particular environmental situation might serve as a niche, but may nevertheless be gone in the next 10,000 years – which is not more than a blink of the eye in geological terms. Therefore, the view that niches correspond to islands of relative stability for generation sequences is one matter, but the question of what will happen to the environment over geological time periods is another. In the following I will therefore discuss the impact of environmental changes on the course of generation sequences during geological time periods.

The emergence of unpredictable environmental changes

Now, what can be said about the behaviour of the environment over geological time periods and the consequences for generation sequences? First of all, the precise behaviour of the environment over long periods of time is highly unpredictable. The dramatic climate variations over the last 100,000 years can serve as a concrete example (see e.g. Lenton T. & Watson A. 2011). As a consequence, over deep time generation sequences are inevitably confronted with multiple environmental changes. In this regard, the magnitude and consequences of the changes would be decisive. For example, they can have such a magnitude that they destroy worldwide ecosystems and their diversity. A reason for such a global catastrophe can be the collision of the earth with a big meteorite. The consequences of this kind of event are radical and profound environmental changes over many generations, concerning not only the impacted region but the whole planet. Other events, such as big eruption of volcanoes or a beginning ice age, have similar consequences (for the different types of catastrophes and their consequences for evolution see again Lenton T. & Watson A. 2011). On the other hand, not all large scale events need to provoke changes at the global level and may be only dramatic in parts of the globe. A respective example would be a catastrophic drought concerning mainly the southern hemisphere. In this regard it should be mentioned that the distribution and magnitude of corresponding environmental changes or catastrophes can be described by a so-called power-law. This says that respective events of different magnitudes can happen at any unpredictable moment. It further says that dramatic large-scale events occur less frequently than smaller ones, in the sense that the magnitudes of respective events show an exponential decrease, if plotted against their frequency (Gribbin J. 2004).

The paleontological record of fossils documents the impact of catastrophic, large-scale events on the evolution of life. Well known in this respect are the so-called "big five", which were all followed by a dramatic mass extinction. In short, over deep time catastrophes of different magnitude, and the environmental changes provoked thereby, are inevitable. If they are on a smaller scale, they may only have limited consequences, if they are large they may wipe out existing ecosystems worldwide (for catastrophes and mass extinctions during biological evolution see e.g. Courtillot V. 1999 or Hallam A. and Wignall P.B. 1997).

The different forms of environmental stress

If large-scale environmental changes or catastrophes happen inevitably over deep time and destroy existing ecosystems, then they must influence biological evolution and thus complexity increase. Therefore I want to discuss the impact of corresponding events on the probability of multiplication of generation sequences.

From the generation sequence point of view, environmental changes can have in principle two consequences. Firstly, the change can affect the niche situation somehow, so that many generation sequences will stop, but not all. As previously said, the probability of multiplication would be influenced by the frequency of negative generation sequences stopping events. Corresponding environmental changes would therefore be part of these negative events. In other words, general environmental stress that does not completely wipe out a niche, and therefore does not stop all respective generation sequences, would cause what Darwin called natural selection. The other possibility is that of changes or catastrophes which destroy existing ecosystems, followed by the extinction of many species. With regard to the following discussion this second kind of event is important, because it opens the way for new ecosystems with a new diversity.

The differentiation between general environmental stress and ecosystem destroying large scale changes or catastrophes is important with regard to the definition of niches. General environmental stress would be part of the niche conditions. On the other hand, a niche has to provide sufficiently stabile or reliable conditions for generation sequences and a corresponding stress can therefore not be too frequent or strong. In contrast, large-scale changes or catastrophes cause the disappearance of established ecosystems. Their frequency must therefore be relatively low so that the likelihood for a corresponding event to occur in the next generation can normally be neglected. This means that general environmental stress would be relevant for the probability of multiplication, but the rare large-scale catastrophic events normally not.

A good example to illustrate this point is the comparison of *Home sapiens* with archaic prokaryotes. Corresponding to the low complexity of the latter, they own a relatively low probability of multiplication. This means the likelihood that an individual prokaryotic generation sequence gets hit by general environmental stress is very high. On the other hand, the archaic forms have existed most probably for about three billion years, which means their generation sequences have stretched over this long time without these organisms showing substantial changes. A reason for this could be that their niches correspond to situations such as hydrothermal vents on the deep sea floor, which have been present for billions of years and were not destroyed by the catastrophes that happened on earth over the last three billion years (for evolution's first three billion years see Knoll A.H. 2003).

In comparison, the difference is striking with regard to the generation sequences that lead to higher organisms, such as *Homo sapiens*. While the latter extend over the same long period of time, as those of today's archaic prokaryotes, the individual forms carrying these sequences have dramatically changed. And so have the corresponding niches. This may be related to the fact that the generation sequences towards *Home sapiens* had to deal with all large-scale changes or catastrophes during earth history, such as the big five mentioned before. Obviously this raises the suspicion that the confrontation of generation sequences with large-scale changes or catastrophes corresponds to a necessary catalyst for biological complexity increase. In the following I will deliver the arguments as to why exactly this is the case. This means that the destruction of established ecosystems by large-scale changes or catastrophes would be central for the complexity increase towards a mounting probability of multiplication.

In summary, over deep time generation sequences face the risk of being confronted with unpredictable catastrophic events of different magnitudes. The overwhelming majority are of low scale and do not destroy whole ecosystems. Some rare large-scale events, however, destroy established ecosystems, either on a regional or global level. Their traces are visible as mass-extinctions in the paleontological record (Courtillot V. 1999 or Hallam A. and Wignall P.B. 1997) and in the following I will discuss their consequences for biological complexity increase.

Complexity increase and ecosystems

Before starting the discussion about the impact of large-scale changes or catastrophes on evolution, I will summarize first the insights gained hitherto about biological complexity increase. It was deduced that the complexity increase of life is directed towards a rising probability of multiplication. This happens through the emergence of new phylogenetic clusters that are characterized by an on average higher probability of multiplication. The latter is the result of an altered dependence on essential niche factors or conditions that causes a strengthening of a weak point in the respective lifecycle at which most of the corresponding generation sequences get stopped. As a consequence, the individual generation sequence of more complex forms owns an on average higher likelihood to continue over time and thus a rising probability of multiplication. The way in which this happens is via dual selection of key innovations. This means that it is the consecutive action of mutual selection and selection on fitness that would produce the qualities which differentiate the phylogenetic clusters.

These insights about complexity increase allow the following conclusion: on their way through deep time some generation sequences become repeatedly the object of a process which drives the emergence of key innovations and a subsequent increase in their average probability of multiplication. In the following I will deliver arguments as to why exactly this is the case. To express the inevitability of this process, I will call it the law-like process of complexity increase.

89

With regard to the question of what this law-like process could look like, let me recall what has been said about key innovations. The acquisition of a key innovation within an established ecosystem would allow the concerned form to occupy a new kind of niche. As said, this would be possible because key innovations alter the dependence on essential niche factors. On the other hand it was also said that within established ecosystems, newly emerging key innovation-owning forms should play a minor role. The reason is that the key innovation does not mean ousting other forms from their established niches. It would rather allow a specialization on certain conditions not suited for other forms, thus the occupation of particular niches. With the restriction to specific niche conditions, it is also difficult to imagine how these key innovation-owning forms could start a development from quantity to quality that would be necessary for a rising probability of multiplication. Now, if this is the case it must have the visible consequence that the complexity increase during evolution is in a kind of standby mode in established ecosystems. And indeed, exactly this is the case – see below.

In the following, I will identify the details of the law-like process of complexity increase. For this, I will first connect the insights about generation sequences with the behaviour of the environment over deep time.

Large-scale environmental changes or catastrophes drive complexity increase

Now, what can cause a situation so that the development towards a rising probability of multiplication occurs? Here, the behaviour of the environment over deep time becomes important, because established ecosystems do not exist forever. Sooner or later they will disappear and new systems with a different profile of forms can emerge. A well-known example is the big meteorite that wiped out the dinosaurs about 65 million years ago and opened the way for the age of mammals (see e.g. Lenton T. & Watson A. 2011). After large-scale catastrophes of this kind, the so-called founder forms are obviously decisive for what the biological diversity in the new ecosystems will look like. Consequently it can be expected that if particular forms have a relative higher likelihood to survive, then the nature of the newly emerging ecosystems should strongly be influenced by these founder forms. With regard to complexity increase this means that the destruction of established ecosystems by large-scale changes or catastrophes can fundamentally change the situation for key innovation-owning forms. In the aftermath they might play a much more dominant role that allows eventually a development towards a rising probability of multiplication.

If these assumptions are correct, it means that the following steps would happen during the course of evolution: firstly, forms with new key innovation emerge within established ecosystems, where they play a kind of specialist role. This has the consequence that as long as the established ecosystems exist, a further complexity increase is blocked. Secondly, the complexity increase can eventually continue in the aftermath of a catastrophic large-scale event. By wiping out many of the hitherto dominant forms, the destruction of the established ecosystems overcomes the stagnation. Hence, it opens the way for a new biodiversity of key innovation-owning forms, characterized by an on average higher probability of multiplication. As the reader will see, these steps are at the centre of the law-like process of complexity increase.

Now, if above scenario corresponds to reality, it must become visible in evolution. And actually this is the case. Let's take the evolutionary history of mammals. After their appearance within established ecosystems dominated by dinosaurs, they developed into a relatively limited diversity of small-sized animals, which were adapted to special niche conditions, thanks to their 'mammalness' (Conway Morris S. 2003). Very probably they needed also a relatively high multiplication rate, in order to survive in a dinosaur- dominated world. As long as the established ecosystems were present, this situation did not change fundamentally and mammals did not show substantial changes during the long reign of dinosaurs. But also the dinosaurs did not change substantially over very long periods of time (again Conway Morris S. 2003). This is exactly what can be expected, if the above argument about the standby mode of complexity increase in established ecosystems is correct. The situation changed when about 65 million years ago a meteorite destroyed the established ecosystems. The relatively limited number of small-sized mammals then evolved into a new, manifold diversity (Lenton T. & Watson A. 2011). And not only this, the mammals were also able to continue the biological complexity increase that resulted eventually in the rise of *Homo sapiens*.

The same arguments can be made for the transition of ecosystems dominated by gymnosperms to those dominated by angiosperms, the modern flowering plants. The first flowering plants known to exist are from 130-136 million years ago, in a time when the flora was dominated by gymnosperms (Soltis D.E. et al. 2008). It is suggested that angiosperms originated as specialists, adapted to dark, damp, frequently disturbed areas. Their restricted initial significance probably only changed after the global climate became much colder in the so-called Cretaceous Period. The climate change provoked mass extinctions and the disappearance of the established ecosystems. This cleared the way for angiosperms, which adapted to the new climate much better. As a result, angiosperms became widespread around 100 million years ago and replaced conifers as the dominant trees around 60-100 million years ago (see e.g. Wing S.L. and Boucher L.D. 1998).

The angiosperm example suggests that big catastrophic events, such as a meteor impact, are not the only reason why established ecosystems disappear and complexity increase can continue. Many ecosystems during earth history were destroyed by climatic changes, such as in the Cretaceous Period (Lenton T. & Watson A. 2011). This is important for the course of complexity increase. For example, in the evolutionary history of *Homo sapiens*, the big meteor impact mentioned was without any doubt very central. But it cannot have been the only relevant change or event. This means that it was a succession of multiple catastrophic events and changes, with some more spectacular or destructive than others, which drove the evolutionary complexity increase from the first primitive mammals to *Homo sapiens* (see again Lenton T. & Watson A. 2011).

The conquest of new habitats

At this point I need to mention an alternative way in which, in certain cases and without the destruction of established ecosystems, key innovation-owning forms can suddenly become dominate in new ecosystems. This I will call the pioneer effect. It has to do with the fact that certain key innovations allow the conquest of new habitats, which were at the corresponding time not accessible for comparable forms without it. Concrete examples are the conquest of terra firma by amphibians, or the conquest of dry habitats by reptiles. The amphibian or reptile specific key innovations allowed, at the corresponding time, the adaptation to a new kind of niche situation, to which comparable forms, but without the respective key innovation, were not able to adapt. This for itself would not be different than in the case of other key innovations. However, in both cases the new niche type corresponded also to a new habitat. As a result, the first amphibians as well as the first reptiles could become the starting point of new ecosystems within the newly conquered environment, without the need for established ecosystems to disappear before. For example, reptiles started to dominate the dry land, thanks to their key innovation characterized by amniotic eggs and waterproof skin. In parallel, large amphibians continued to exist in their traditional, wetter habitats, until they became extinct by the massive Triassic-Jurassic extinction event. This happened about 200 million years ago and opened the way for the long reign of dinosaurs (Lenton T. & Watson A. 2011).

The pioneer effect may also have been relevant for other important events during evolution, such as the diversification of the first unicellular eukaryotic cells or those of multi-cellular organisms. In both cases, the respective key innovations may have allowed the conquest of habitats, not accessible for the other forms at that time – in part 2, I will discuss these steps in evolution in more detail. In summary it can therefore be said that some key innovations may not only provide the

92

capability to occupy a new kind of niche, but allow as a result also the conquest of a new kind of habitat. The respective key innovation-owning pioneers can therefore conquer this new environment, without established ecosystems having to first disappear.

The probability of multiplication and large-scale catastrophes

Let's now search for an explanation of how key innovation-owning forms can eventually raise their probability of multiplication. It was said that the appearance of taxonomic clusters with an on average higher probability of multiplication represents the visible result of biological complexity increase. The increase in the probability of multiplication would document a niche situation with on average relatively more stable or reliable conditions for the individual generation sequence. In view of what has been said so far, the central question is thus: when and why can the key innovation-owning forms start with the required development from quantity to quality? As said, as long as the established ecosystems exist and the corresponding forms are restricted to particular niche situations, this should not happen. The situation can change however, at the moment a large-scale environmental catastrophe or change occurs and suddenly the established ecosystems do no longer exist. Figuratively speaking, a relatively small number of survivors find themselves in an empty world, because many of the hitherto dominant forms became extinct. After the situation is beginning to recover, the surviving forms can take possession of the world and become the starting point for new ecosystems with a new diversity. Consequently, the profile of the new diversity depends on the lucky survivors of the mass extinction (see e.g. Conway Morris S. 2003 or Lenton T. & Watson A. 2011). In case these are not the forms that dominated the ecosystems so far, it can be expected that previously less important forms will now play a much more dominant role.

With this in mind, let's assume that key innovation-owning forms would actually survive a large-scale catastrophe, but not the thus far dominant forms. What would be the consequence? As said, the key innovation-owning forms might become much more dominant, compared to their previous minor role. Without the formerly dominant forms, they are no longer restricted to a specialist role. Now the key innovation-owning forms have the possibility to adapt to all suited niches in the newly developing ecosystems, including ordinary niches. By this I mean niches which existed in a comparable way already before the catastrophe, but where they were occupied by other forms. Here the following becomes decisive, which is central for complexity increase. Compared to the previous situation, the key innovation strengthens in this kind of niche the weak point in the lifecycle, so that the corresponding individual generation sequence has a higher likelihood to continue over time. From the sustainability point of view this must have consequences. Let's take the example of big herbivores. If key innovation-owning forms survive a mass extinction and now become big herbivores, their niches would be similar to those that were previously occupied by comparable forms, but without the key innovation. As is the case for all niches that are at risk of becoming irreversibly exhausted, the respective generation sequences need to keep a sustainable balance with the environment. Under comparable niche situations however, the individual generation sequence of forms with key innovation would have a relatively higher likelihood to continue over time. To be sustainable, the adaptation of key innovation-owning forms to the niches of big herbivores would therefore require a development from quantity to quality. As a result, big herbivores with key innovation would have an on average higher probability of multiplication than the corresponding forms before the mass extinction.

In summary, from the *Interaction Theory* point of view the aftermath of ecosystem-destroying large-scale changes or catastrophes would be a decisive moment for complexity increase during evolution. It would give key innovation-owning forms the possibility to occupy a wide range of niches in the newly emerging ecosystems following the mass extinction. This includes those niches for which the key innovation is not essential. However, since the strengthening of the weak point in the lifecycle causes a relatively more stable or reliable situation for the individual generation sequence, this may result in an unsustainable growth, compared to the previous situation. Consequently, under these ordinary niche situations, the key innovation-owning forms require a development from quantity to quality, in order to stay in a sustainable balance with the environment. This becomes visible in an on average higher probability of multiplication, compared to the situation before the catastrophe.

The sequence to a higher probability of multiplication

Based on what was said, the emergence of a phylogenetic cluster in the course of evolution, which shows an on average higher probability of multiplication, can be summarized in the following sequence of steps:

1. Within established ecosystems, multiplying forms acquire a new key innovation that strengthens a decisive weak point in the lifecycle by overcoming, or at least significantly reducing, the multiplication's dependency on a hitherto essential niche factor or condition with limited reliability.
2. The result is a new, however limited diversity, because the key innovation allows the specialisation on a new kind of niche, but not the ousting of existing forms from their established niches.

3. The adaptation to the new kind of niche does not demand a development from quantity to quality. It may rather require a high multiplication rate, in order to compensate for a strong likelihood that the individual generation sequence is stopped under the special conditions.
4. Over deep time a large-scale environmental change or catastrophe happens inevitably and destroys the established ecosystems. The key innovation-owning forms survive the consequential mass extinction much better than the previously dominant forms, and thus play a central role in the following development of new ecosystems.
5. As a consequence, they become much more dominant, by adapting to a wide variety of niches. This includes also ordinary niche situations, for which the key innovation is not essential, but still causes a relatively more stable or reliable situation for the individual generation sequence.
6. The sustainable adaptation to these niche situations demands therefore a development from quantity to quality. The result is a phylogenetic cluster of key innovation-owning forms with an on average higher probability of multiplication. In the end, this corresponds to a complexity increase relative to the situation before the mass extinction.

This sequence of steps would explain why complexity increase during evolution occurred stepwise and why the biological diversity along the phylogenetic tree of life shows a clustered organization with the most complex forms at the top.

Key innovations and the survival of catastrophic large-scale events

In view of the proposed steps for complexity increase the decisive question is now: why would key innovation-possessing forms, which play a secondary role as specialist in an established ecosystem, survive a large-scale catastrophe better than the comparable forms which dominated the ecological situation thus far? Or to give a concrete example, when about 65 million years ago a large meteorite wiped out the ecosystems dominated by dinosaurs, why did mammals and birds survive and prevail?

Key innovations, such as those that distinguish mammals from reptilians or angiosperms from gymnosperms, were defined as qualities which strengthen a decisive weak point in the lifecycle of the corresponding form, compared to the respective phylogenetic ancestors. The reason for the strengthening is that the respective key innovation makes it possible to overcome, or at least to significantly reduce, the dependence on particular, hitherto essential environmental factors or conditions, which are responsible for many generation sequences getting stopped. For multi-cellular, cell-differentiated organisms the decisive weak point concerns the vulnerable early phases in the lifecycle. As said, in the case of mammals and birds, the respective strengthening happens thanks to a much greater independence of climatic factors. In the case of reptilians, the strengthening is the result of their independence from spawning grounds or in angiosperms by overcoming the dependence on wind as sole means for pollination and seeding. Key innovations have therefore the two important consequences:

1. They allow the multiplication under specific conditions under which comparable forms without the key innovation cannot multiply.
2. Moreover, they cause under ordinary niche conditions, for which the key innovation is principally not essential, a more stable or reliable situation for the individual generation sequence.

For complexity increase during evolution, it would be the combination of both consequences of a key innovation that is decisive.

With regard to the initial question, it makes sense to assume that strong environmental stress, as it is caused by catastrophic changes or events, would hit the weak points in the lifecycle in particular. And this can only mean that the possession or non-possession of a key innovation that strengthens the weak point can suddenly be of general importance for the survival of the generation sequences. Let's take terrestrial vertebrates for instance. Disappearing nutritional resources, because of a dramatic global climate change, will hit adult amphibians, reptiles and mammals in a comparable manner. On the other hand, it is obvious that the impact on the early, more vulnerable phases in the lifecycle can be very different for the three types of organisms. If the climate becomes for instance very dry, the spawning of amphibians is a decisive disadvantage compared to more water-independent kinds of reproduction. Correspondingly, if the climate becomes suddenly cold, the reproduction of warm-blooded mammals and birds can still be possible, in contrast to cold-blooded forms. Thus, a strengthening of the weak point in the lifecycle provides clear benefits under extreme circumstances. This in turn supports the assumption made that the possession or non-possession of key innovations is then decisive when it concerns surviving a catastrophic change or event, because those will hit in particular the weak points in the lifecycle.

Obviously a direct comparison regarding the potential to survive a catastrophe only makes sense for comparable forms. By analogy to what was already discussed in the case of multiplication attempts per time as a measure for the relative complexity, it makes no obvious sense to compare in this regard for example mammals and angiosperms. The direct comparison regarding the prospect of surviving a general mass extinction would therefore require forms for which the main difference lies in the possession or non-possession of a particular key innovation.

In other words, it concerns related phylogenetic groups or clusters, such as dinosaurs and birds or reptiles and mammals etc. In this context it may also be important to keep in mind that the survivors of a catastrophe will have to compete with each other for the very few resources still available. And this competition should be most severe, the more similar the respective lifecycles are.

The destruction of established ecosystems during earth history

Equipped with these insights, let's look again on the event which occurred about 65 million years ago and ended the long reign of the dinosaurs. Before the meteorite hit the earth, the number of early mammal species was relatively low, and their body size stayed relatively small (Lenton T. & Watson A. 2011). This would be in line with the assumption that they were adapted to special niche conditions that may have demanded small-sized animals with a different dependence on certain environmental factors, compared to the dominant dinosaurs. Hence, these specific niche situations would not have been accessible for small-sized dinosaurs. By comparison, the mammal-specific key innovations were of no decisive relevance for the niches occupied by dinosaurs. Similar arguments can be made for birds, the closest living relatives of dinosaurs. They too would have played more a secondary kind of specialist role, before the catastrophic event destroyed the established ecosystems (for the evolution of birds see e.g. Feduccia A. 1996).

In contrast to the early mammals and birds, many dinosaurs became very big during their evolution. As mentioned earlier, when species become bigger, their multiplication rate goes normally down. The respective body size increase of dinosaurs can therefore be seen as a trend towards a gradually mounting probability of multiplication, which would have been driven by selection on fitness. A possible explanation could be a fierce competition for breeding grounds, for which a bigger body size can provide a decisive advantage and so be favoured by selection on fitness. In the aftermath of the meteorite catastrophe however, the big body size could have suddenly been a decisive disadvantage, because no big terrestrial animals survived (Hallam A. and Wignall P.B. 1997). Here could be a reason why dinosaurs disappeared completely. They would have been too big.

The key innovation of mammals, with qualities such as placenta, mammary gland, regulated thermo-metabolism and insulating fur, are of direct or indirect relevance for the early, vulnerable life-stages. Similarly, birds own a regulated thermo-metabolism plus feathers for thermal insulation, which allows them to brood their eggs. As long as the world was governed by cold-blooded and egg-laying dinosaurs, however, these qualities would have been only beneficial under certain circumstances. This would have changed in the aftermath of the meteorite impact, which caused massive climate changes and temperature drops (Courtillot V. 1999). It is easy to imagine that under these dramatic conditions, the mammal- and bird-specific qualities provided suddenly a general and decisive advantage. As a consequence mammals and birds prevailed in the aftermath of the catastrophe and became the origin of a new diversity and new ecosystems.

In contrast to the total extermination of dinosaurs, other similar events during evolution had a less dramatic result. For example, gymnosperms and amphibians lost their dominance after the rise of angiosperms respectively reptiles, but they did not vanish (see e.g. Soltis D.E. et al. 2008 and Hallam A. and Wignall P.B. 1997). This shows that all possible types of organisms can potentially survive catastrophes, and not only forms with the "latest" key innovation.

Large-scale catastrophes during earth's history and complexity increase

The assumption that catastrophic large-scale changes or events drive biological complexity increase should be visible in evolution. This means that life should have become more complex in the aftermath of the global catastrophes, which are known from earth history. In addition, in the new ecosystems which developed after the respective mass extinctions, the previously dominant forms should be substituted by new forms that played previously a minor role. In the following is a brief summary of what I found in this regard about some of the known mass extinctions during earth history (for more details see e.g. Hallam A. and Wignall P.B. 1997 or Courtillot V. 1999):

Cretaceous-Paleogene extinction event:

- This event, which destroyed about 66 million years ago the dinosaur-dominated ecosystems, was already discussed before. Therefore I need only add that the eventual result of this catastrophic event was not only the rise of mammals and birds; it drove also the further auto-complexification of life toward brain complexity and the rise of *Homo sapiens*.

Triassic-Jurassic extinction event:

- In line with the assumption made are the facts that in the aftermath of this massive catastrophic event about 200 million years ago, life on earth became more complex than before and the dominant forms were replaced by previously less important forms. Concretely, it caused for example the extinction of most non-dinosaurian archosaurs, most therapsids, and most of the large amphibians, thus opening the way for the long reign of the dinosaurs. From the presented scenario for complexity increase it follows that the dinosaurs must have owned a key innovation that strengthened a weak point in the lifecycle, compared to the previously dominant forms. With regard to amphibians it is obvious and was mentioned in detail in the text. However, regarding archosaurs and therapsids I must acknowledge that I don't know. Nevertheless, from what was said it follows that it should be related to the vulnerable early phases in the individual lifecycle. Thus, it can be speculated that dinosaurs had owned a particular quality which made directly or indirectly either their eggs, their youth phase or both less vulnerable, in comparison to the non-dinosaurian form. However, I could not find corresponding references in the literature.

Permian-Triassic transition event:

- This event is also called the "great dying", happening about 252 million years ago. It corresponds to the largest known mass extinction during earth history. But again, also this great dying did not stop life from becoming more complex, even if the full recovery took millions of years. In the end, it was this huge catastrophe that opened the way for a more complex diversity of terrestrial vertebrates. In line with the proposed scenario for complexity increase, the catastrophe ended the primacy of the previously dominant synapsids or theropsids, and created the opportunity for archosaurs to become ascendant. As before, I must acknowledge that it is not possible to say what kind of key innovation helped archosaurs to survive the event better than the synapsids. Again, it should be related to the vulnerable phases in the lifecycle. Also in this case I could not find corresponding references in the literature.

Cambrian explosion:

- The so-called Cambrian explosion about 542 million years ago describes a relatively short evolutionary period in the Cambrian Period, during which most major animal phyla appeared (Gould S. J. 2002). Before that, most organisms were simple, mostly composed of individual cells. The Cambrian explosion represents therefore a decisive step for complexity increase during evolution. In its aftermath more complex organisms appeared and the diversity of life began to resemble that of today. From the view of the identified scenario for complexity increase it should therefore be expected that also this event can be linked to drastic change in the environment. And indeed, indications for corresponding environmental changes exist which could have opened the possibility for moderately complex animals to become dominant and to diversify in a relatively short period (Lenton T. & Watson A. 2011).

In the following, I will now come to the central point of *Interaction Theory*, the formulation of the law-like process of complexity increase.

1.12. The law-like process of complexity increase

In this section, I will introduce a law-like process of complexity increase to describe the principles behind the biological auto-complexification during evolution. The process explains how in some generation sequences of life the same series of events repeats over and over again and thus leads to increasingly complex biological forms. And because of its inevitability, the process would be law-like.

The law-like process is based on the insights gained about the progression of generation sequences through deep time and the associated meaning of biological complexity – a summary of the main insights can be found in the annex. This results in the following conclusion about biological complexity increases:

- Biological evolution is subject to a law-like process of complexity increase that adds progressively new groups to the phylogenetic tree of life. Each of these groups owns a respective key innovation that results, in the aftermath of a mass extinction, in a higher average probability of multiplication of the corresponding generation sequences.

The law-like process of complexity increase is at the heart of *Interaction Theory*. It consists of two parts. Firstly, the acquisition of a new key innovation that strengthens a decisive weak point in the lifecycle and allows the respective generation sequences to adapt to a new kind of niche within established ecosystems. Secondly, thanks to the less vulnerable lifecycle, the corresponding generation sequences are not as easily stopped by a sudden, catastrophic event or change as those of comparable form without the key innovation. As a consequence, the key innovation-owning forms can thrive in the aftermath of a mass extinction caused by an ecosystems destroying environmental catastrophe. They can now form a new, much more varied diversity of key innovation-owning forms, which shows an on average higher probability of multiplication, and hence corresponds to an increase in biological complexity. This succession of steps is illustrated in flowchart 1. Each round produces generation sequences with a progressively higher average probability of multiplication, before the whole process begins again. Over deep time, the repeating rounds cause a clustered organisation of the biological diversity that stretches from low to high complex forms, as is visible in the phylogenetic tree of life.

The law-like process of complexity increase in detail

The law-like process starts with the development of ecosystems in which the generation sequences adapt to the available niches, resulting in a corresponding diversity of forms (see flowchart 1 page 172). Over time the ecosystems become established, in the sense that the available niches are largely occupied. As a consequence of the respective niche adaptations, the probability of multiplication may increase or decrease for some generation sequences. This would be the result of an adaptation to a particular niche situation driven by a selection on fitness.

In niches at risk of becoming irreversibly exhausted by exponential growth, the generation sequences require reproductive interactions, such as sexual reproduction, for sustainability reasons. Thus, the corresponding forms are also exposed to a mutual selection that is independent of selection on fitness. As a result, the dually selected forms in this type of niche are characterized by the possession of qualities that are either a consequence of selection on fitness or of mutual selection (step 2). Likewise, the forms in niches tolerating exponential growth may acquire qualities that are not the result of selection on fitness, but a side-effect of symbiosis. In both cases the interaction- or symbiosis-related quality can become relevant for the fitness of the corresponding form and therefore an object for selection on fitness (step 3). As a consequence, the eventual outcome is a product of dual selection. This sets them apart from qualities which are the exclusive product of selection on fitness, without the need for an upfront investment, either by mutual selection or symbiosis. The origin by dual selection has the consequence that if a corresponding quality shows a positive impact on the environmental adaptation, it can lead to a new phylogenetic group or cluster. The members of a respective group would be characterized by the possession of particular dually selected qualities which could not be acquired by environmental adaptation. As mentioned, examples would be taxonomic groups such as cat- or dog-like species.

Over time, dual selection may also cause the emergence of a key innovation (step 4). These are qualities that make the multiplication independent of a hitherto essential niche factor or condition with limited reliability, which weighs on the probability of multiplication of the respective generation sequences. This means that a key innovation strengthens a decisive weak point in the lifecycle of the respective form where most of its generation sequences stop. The consequences are therefore twofold. Firstly, the altered dependence on niche factors or conditions

101

makes it possible for the key innovation-owning forms to multiply under conditions which are not possible for comparable forms without the key innovation. Thus, they are able to adapt to a new kind of niche (step 5). By adapting to corresponding niches, the key innovation-owning form becomes the origin of a new phylogenetic group, characterized by the possession of the respective key innovation. Secondly under more common or ordinary niche conditions, the key innovation can increase the likelihood for the individual generation sequence to continue over time, and therefore raise the probability of multiplication. However, within the established ecosystems these ordinary niches are occupied by the dominant diversity of forms without key innovation. And because, a potentially higher probability of multiplication does not represent an essential benefit in the here and now, the key innovation-owning forms cannot oust the dominant forms from their established niches. They are therefore restricted to particular situations where their altered dependence on niche factors or conditions represents a tangible advantage. Thus, they can only play a minor or specialist role, as long as the established ecosystems exist.

Over deep time, a catastrophic large-scale change or event happens inevitably, causing the destruction of the established ecosystems, followed by a mass extinction (step 6). As a consequence of their altered dependence on environmental factors or conditions and the resulting more robust lifecycle, the generation sequences of key innovation-owning forms survive better than those of the hitherto dominant forms. For this reason, the key innovation-owning forms can play a dominant role in the new ecosystems which develop after the disaster. Here they can also adapt to those niche situations previously occupied by the formerly dominant forms. For their sustainable adaptation to these ordinary niche situations, for which an altered dependence on environmental factors or conditions is not essential, they require, however, a development from quantity to quality (step 7). The reason is that here their generation sequences become on average less frequently stopped, due to their less sensitive lifecycle.

The new ecosystems in which the key innovation-owning forms play a dominant role become established over time. Here, the development from quantity to quality that occurred under multiple niche situations results in the following consequence. The generation sequences of the now dominant, key innovation-owning forms show an on average higher probability of multiplication than the previously dominant forms (step 8). Thus a complexity increase took place compared to the situation before the large-scale catastrophe.

In the established ecosystems, the law-like process of complexity increase can then restart again, so that with each round more and more complex forms emerge, of which the individual generation sequence owns an increasingly greater likelihood of continuing over time. And because the process is driven by the dual selection of key innovations, it results in a clustered organisation of the phylogenetic tree of life.

Up to this point, I examined the question of why and how the complexity of multiplying forms increases during evolution. To do so, I tried to clarify the way in which generation sequences can travel through deep time, in an environment with exhaustible resources and the risk of sudden catastrophic events or changes. The outcome is the *Interaction Theory* with the law-like process of complexity increase in its centre.

In the following part of the book, I will now deduce what can be said about the course of evolution as a consequence of the law-like process of complexity increase. Afterwards I will compare the result with the known biological facts. It will turn out that the predictions following from the law-like process are indeed very much in line with what has happened during biological evolution.

Part 2: The course of complexity increase during evolution

2.1. The journey of generation sequences through deep time

In the previous part I introduced *Interaction Theory* to describe the complexity increase of multiplying forms during evolution. In contrast to the currently prevailing understanding, the theory puts the generation sequence of life and a law-like process of complexity increase into the centre. At this point, the obvious question is, how well does *Interaction Theory* describe the reality of biological evolution? And in particular, whether it describes the evolution of life better than conventional evolutionary theory? In the rest of the book I will provide arguments why this is indeed the case.

In this context it must be considered that a direct experimental proof for a theory concerning historical events is not possible (see e.g. Lloyd E.A. 1994). But a respective theory must fulfil the plausibility criteria, in the sense that its results must be in accordance with what is known about the evolution of life. In addition, the theoretical model should allow logical conclusions, which can then be compared with reality (Fontana W. and Buss L.W. 1994). Hence, if complexity increase during biological evolution follows the principles of *Interaction Theory*, the course of evolution must be in accordance with the law-like process towards a rising probability of multiplication. And I will carry out precisely such a plausibility check in the remaining pages of this book. In particular I will try to show that the fundamental principles of *Interaction Theory* allow meaningful predictions about the course of biological evolution and that these are in line with what is known about the history of life on our planet.

Rewinding the tape

The evolution of life on earth with its resulting biological complexity has stretched over a period of probably 3 to 4 billion years. It started with relatively simple molecules, then went to prokaryotes then eukaryotes and eventually to highly complex forms such as *Homo sapiens* (for evolution and complexity see e.g. Heylighen F. 1996). From the *Interaction Theory* point of view this evolution had to follow the path of a continuously rising probability of multiplication of some generation sequences and thus to increasingly complex biological forms. This fundamental principle, together with the insight that life had to start under irreversibly exhaustible conditions (see below), allows predictions about the journey of generation sequences through deep time. So, it can be deduced that the whole evolution from molecules to *Homo sapiens* can be divided into four distinct phases of complexity increase. The phases can be differentiated by the way in which the probability of multiplication of the respective generation sequences was able to rise. Thus, after the first, primitive multiplying molecules emerged, the law-like process of complexity increase conducted their generation sequences predictably through four phases towards an intelligent being. Obviously this view stands in sharp contrast to established evolutionary thinking and in the following I will deliver the arguments to support it.

Now, if the evolution of life becomes predictable, in the sense that it has to follow the path towards a rising probability of multiplication, it raises the question: if the history of life were to happen again, can the result be exactly the same? From the *Interaction Theory* point of view the answer is yes and no. By this I mean the following: if the law-like process of complexity increase gets restarted under identical conditions, it should follow the same way on the macro-level. Hence if sufficient time is given, evolution would reproduce for example solitary multiplying forms comparable to prokaryotes. In addition, it should also reproduce an intelligent being because intelligence is a quality that can strongly increase the probability of multiplication (see below). Therefore if we can rewind the tape on earth, as Stephan Gould has phrased it, the development of a phylogenetic tree, probably not so much different from ours, should occur again.

On the other hand, on the level of the individual forms and species the outcome cannot be exactly the same. For example, even if the law-like process of complexity increase results again in highly social and intelligent species, the question of what these forms would exactly look like cannot be answered. The situation is comparable to the question if the geological forces which have shaped the earth surface would reproduce the identical outcome. Also in this case it is not to be expected that for example the coastline of Norway would be remodelled exactly as it is today. The reason is that the details of the corresponding process are influenced by unpredictable events and circumstances. For the law-like process of biological complexity increase it would be analogous. It depends on events and circumstances which are unpredictable, such as the precise outcome of mutual selection or the exact time and magnitude of an environmental catastrophe. For this reason the exact look of the forms and species, as we know them today, would also be unpredictable. But does this mean the biological world would look very different, if we rewind the tape? In this regard, I would like to quote the already mentioned book of Simon Conway Morris about the role of convergence in evolution (Conway Morris S. 2003). Here, the interested reader can find an answer to this question in form of a quotation of Robert Bieri: "... they won't be spheres, pyramids, cubes, or pancakes. In all probability they will look an awful lot like us."

Key events during the journey of the generation sequences of life

In the following I will discuss several key events in the history of life, such as the emergence of prokaryotes or eukaryotes. For this I will examine each of the key events under the aspect of the law-like process of complexity increase and in particular considering the following questions:

- Did the probability of multiplication rise at the respective point of evolution?
- Does a weak point in the respective lifecycle exist that may be decisive for the probability of multiplication of the generation sequences?
- By what kind of key innovation can the weak point be strengthened, as a precondition for a rising probability of multiplication?
- What type of dual selection can be at the origin of the respective key innovation?
- What type of large-scale environmental event or change can drive the breakthrough of the respective key innovation-owning forms?

Obviously, the answers to these questions have to make sense with regard to the known facts and provide meaningful insights about the course of evolution.

To avoid any misunderstanding, it is not the intention to imply that with *Interaction Theory* it is possible to know exactly what has happened in the history of life. As said, the assumptions that can be made on the basis of the law-like process are related to the macro-level. Here, the analogy with assumptions about the flow of a river from its source to the ocean may be helpful. The basic rule that water flows downhill makes it possible to explain in retrospect why a particular river took a certain and not another direction. However, it is not possible to predict each little river bend since not every little detail of the corresponding landscape at the respective time is known. With regard to biological evolution it would be similar. The knowledge of the details and circumstances about a particular period of evolution is in general very limited. Therefore it can for instance be difficult to understand precisely why the possession of a certain key innovation resulted in a better survival after a large-scale catastrophe, or what the consequences of this catastrophe exactly looked like. Despite this kind of restriction, it will become clear that the search for answers on the above questions will significantly improve the understanding of why the complexity increase during evolution took certain directions.

2.2. From molecules to prokaryotes

There are good reasons for assuming that on earth the generation sequences of life have been on their way for more than 3.5 billion years (see e.g. Doolittle W.F. 2000). The starting point would correspond to the beginning of the evolution of life. In the following I will follow the course of biological evolution, by asking what exactly has happened to the generation sequences on their long journey through deep time. As mentioned above, the journey of generation sequences can be divided into distinct phases which are different in the manner in which the probability of multiplication of the respective generation sequences was able to rise. In the following I will start with the first of these phases, which stretched from molecules with the capability to multiply to prokaryotes.

The kick-off of generation sequences needs interaction communities

Good arguments exist as to why evolution had to start with relatively simple molecules with the capability to multiply – or be multiplied, would probably be more correct (see e.g. Kauffman S. 1995 and 2003). Over time these molecules became structurally more complex, gained new catalytic capabilities and formed, together with membrane-like structures, the first 'cellular forms', which got eventually transformed into prokaryotes with outstanding biochemical power. While there is little dispute about the molecular origin of life, it is neither known what these molecules exactly looked like, nor what type of conditions made their emergence possible (see e.g. Maynard Smith J. and Szathmáry E. 1995). Nevertheless, in the following I will try to gain new insights about this early phase of life, by approaching it from the *Interaction Theory* point of view.

In the first part of this book it was already mentioned that the entities at the beginning of life should have been molecular interaction communities. This would correspond to the chemical networks, as for example proposed by Stuart Kaufmann, following from the reasoning about the probability of events that could cause the emergence of life (see e.g. Kaufmann 2003). The new and additional argument of *Interaction Theory* is that respective molecules can only then become the origin of life, if they kick off generation sequences that can acquire a mounting probability of multiplication via the law-like process of complexity increase.

Molecular interaction communities, respectively chemical networks, stand in contrast to the hypothesis that the beginning of evolution occurred in a so-called RNA world (for the latter see e.g. De Duve C. 1995). This means that self-replicating RNA molecules would have been the entities that started biological evolution. However, such molecules would correspond to solitary multiplying forms and consequently fall into the trap of an overall growth contest. In other words, they would experience the same destiny as the RNA in the mentioned Spiegelman trial. Their growth contest would favour the variants that multiply the fastest by reducing the molecular structure to the minimum. If it is further assumed that generation sequences can only start with forms that own a sufficient probability of multiplication, then it means that self-replicating and therefore solitary multiplying molecules are anyhow bad candidates. As discussed in the first part, in a world with exhaustible resources their multiplication would inevitably become trapped in the equilibrium between formation and degradation, which represents a dead end for generation sequences.

In the first part it was also said that in contrast to solitary multiplying forms, reproductive interaction communities can discriminate between variants with different interaction competence. This discrimination or mutual selection can also occur in an overall equilibrium situation between formation and degradation (compare illustration 1). In other words, thanks to their mutual selection on interaction competence, molecular interaction communities would not simply accumulate the fastest replicating variant. They are rather exposed to a dual selection, which corresponds to a mandatory condition for the law-like process of complexity increase to get started.

What could the chemical networks or molecular interaction communities at the beginning of life have looked like? The probably simplest version of a network would be one in which each member is directly involved in the multiplication of at least one other member. As already mentioned the basic structure of the cellular metabolism supports this. All cellular activities are based on a reproductive interaction between DNA, RNA and protein. The three macromolecules need each other to become multiplied, as is visible in the transcription and translation process of the cellular metabolism (see e.g. Campbell N.A. et al. 2008). Thus, the mutual dependence between DNA, RNA and protein supports the argument that evolution and life is founded on reproductive interactions. On the other hand, early interaction communities between DNA, RNA and protein would not be good candidates to get the generation sequences of life started. The reason is that it is difficult to imagine that these sophisticated macromolecules can exist under prebiotic conditions (see e.g. de Duve C. 1995 or Shapiro R. 2007). It therefore makes more sense to assume that the early interaction communities were formed by simpler precursor molecules. During the first phase of evolution these precursors could then be transformed into the familiar DNA, RNA and proteins. The assumption that evolution started with simpler molecules than the cellular macromolecules is in line with general thinking about the beginning of evolution, even if it is unclear what those precursors had looked like (Peretó J. 2005).

The fundamental importance of reproductive interactions already in the very early phase of evolution is also documented by the reproductive interaction between cellular membranes and the rest of the cell. Also here, both parties can be seen as interaction partners that are in a mutual dependence regarding their multiplication. In other words, cells do not synthesise membranes *de novo*; they only multiply already existing membranes (see e.g. Watson H. 2015). Again this can be seen as the relic of the necessity that life had to start with reproductive interaction, in this case between chemical networks and phospholipid vesicles. In other words, the chemical networks at the beginning of life would have facilitated the synthesis of certain phospholipid molecules and helped in this way the respective vesicles to grow. In exchange, the latter would have allowed the members of a network to attach and not to get diluted. The resulting mutual dependence could then have been the basis for the development of cellular structures. In conclusion, the fundament of the cellular metabolism as well as the basic structure of the cell can be seen as the visible sign that life and evolution started with reproductive interactions, as is asserted by *Interaction Theory*.

The weak point of generation sequences at the beginning of evolution

In this section I will discuss whether a general weak point for the probability of multiplication of generation sequences at the beginning of life can be identified. As the reader will see, the answer is yes and allows conclusions about the development of the first chemical networks into prokaryotic forms – compare here also the section about the probability of multiplication at the beginning of evolution in the first part.

As previously said, from the *Interaction Theory* point of view it is assumed that the multiplying forms at the beginning of life consisted of relatively simple molecules which were multiplying via a reproductive interaction. In order for these molecules to be able to start the generation sequences of life, it required suitable environmental conditions. This means that it is not enough that the corresponding conditions made possible the multiplication as such. They must have been also available with a sufficient reliability or stability so that the multiplication was able to continue over time. Now, what can corresponding conditions look like? In this regard it can be said that they must have provided all resources and factors necessary for the respective multiplication. Here it makes sense to suppose that the first molecular interaction communities would not have had much more catalytic capabilities than to enable their mutual formation. At the beginning this meant probably not much more than that the molecules catalysed mutually the linking together of particular building blocks. At the same time, these molecules would not have owned the catalytic power to synthesise necessary building blocks from simple and very reliably available precursor substances, such as CO_2 or methane.

As a consequence of such limited catalytic capabilities, the multiplication at the beginning of life would have required the availability of highly specific conditions and factors. In other words, the early molecular interaction communities would have been dependent on highly specific environmental conditions, characterized by the simultaneous presence of a variety of different factors and resources at the same time and same place. Beside the respective particular building blocks, this would include multiple other factors, such as specific energy-rich substances to deliver the necessary chemical energy. Obviously, environmental situations, consisting of a specific mix of factors and conditions would bear a high risk that not all essential components are permanently present. Hence, the generation sequences at the beginning of life would have been exposed to a very high and permanent risk of getting stopped, because the required sophisticated mix of factors and conditions was not reliably available. In summary, the generation sequences caused by molecular interaction communities at the beginning of evolution were under the constant threat of getting stopped by the low reliability of the environmental conditions they needed for their multiplication. The high specificity of these conditions can therefore be seen as a decisive weak point for the early generation sequences.

In comparison, prokaryotes can metabolise almost any substance, thanks to their exceptional biochemical power (Margulis L 1993). This capability together with their ability to form endospores enables prokaryotes to occupy niches which can regenerate after an exhaustion or destruction by exponential growth. In this regard autotrophic forms, such as methanogens or cyanobacteria, are of particular importance. They can grow on simple substances and factors, such as methane or carbon-dioxide and sunlight, which are very reliably available and regenerate independent of their consumption. Once corresponding autotrophic prokaryotes were present, they provided the basis for additional regenerating niches for heterotrophic forms. The autotrophic and heterotrophic forms can then build together corresponding ecosystems. The fact that cyanobacteria and methanogens still exist today and that these forms can probably be traced back in palaeontology over 3.5 billion of years (Knoll A.H. 2003) shows that their respective niches are present over time with an exceptionally high reliability. In the end this means that prokaryotic forms did definitively overcome the identified weak point of the generation sequences at the beginning of evolution.

As a consequence of these differences between molecular interaction communities as the first multiplying entities on the one hand and prokaryotes on the other, it can be concluded that the first phase of evolution stretched from a multiplication depending on highly specific conditions and factors to multiplying forms growing on almost anything. The highly specific requirements at the beginning would have caused a very low reliability of the corresponding conditions that were necessary to sustain the multiplication over time. With prokaryotes it ended eventually with forms able to occupy regenerating niches that are characterized by a very high corresponding reliability, because of relatively simple and easily available substances and factors. Interestingly this development goes hand in hand with the transformation of structurally relatively simple entities, the molecular interaction communities, into comparably much more complex prokaryotic cells. This gives the impression that the start of evolution requires the transfer of "external" complexity into an "internal" complexity, in order to enable a more reliable situation for the generation sequences (complexity expresses here the relative amount of information needed to describe a form or situation in detail).

The probability of multiplication in the first phase of evolution

For the individual generation sequence at the beginning of evolution, the average likelihood to proceed over time must have been determined by the discussed dependence on very specific conditions and factors for the multiplication. The need for particular environmental situations which are easily disturbed and thus of low reliability can therefore be seen as the very first, decisive weak point for the probability of multiplication in the evolution of life. As a consequence of the central dogma of *Interaction Theory*, namely that biological complexity increase goes along with a rising probability of multiplication, it can be assumed that the probability of multiplication at the beginning of all generation sequences was at its lowest point, relative to the rest of evolution. And this in turn means that the probability of multiplication would have continuously increased on the way to prokaryotic-like cells, by adapting to successively less specific conditions with a progressively growing reliability, so that the multiplication was better sustained over time.

This development could have become possible via the stepwise acquisition of qualities allowing the usage of resources and factors which were progressively more reliably available. An example would be the acquisition of a quality that makes it possible to transfer an easily available precursor into a necessary building block for the multiplication, compared to the original dependence on the building block as such. A corresponding development would come to an end when it results eventually in the ability to use resources and factors, such as carbon dioxide and sunlight, which are very easily available and in this sense highly unspecific. Thus, the first part of evolution would have been characterized by the step-wise acquisition of qualities which allowed a multiplication under environmental conditions that are increasingly more reliably available. And conditions that own in this sense a very high reliability are those that can be easily found in the environment and are also not irreversibly exhaustible or destroyable by growth, such as the niches occupied by autotrophic prokaryotes. The development towards prokaryotes with exceptional biochemical power and capable to grow on all sorts of substances would therefore be the result of an evolution towards a rising probability of multiplication. The autotrophic, regenerating niches of prokaryotes can be seen as a kind of attractor for the generation sequences at the beginning of life, after the first molecular interaction communities appeared.

The above argumentation also fits with the fact that life is based on polymeric macromolecules such as DNA, RNA and protein, and not on non-polymeric structures. The environmental conditions for the multiplication of the very first molecular interaction communities had to fulfil fewer conditions, if they consisted of polymeric structures, compared to a situation with non-polymeric molecules. The reason is that polymers consist of monomers and the likelihood that different monomers of a certain type are given at a certain place should in general be higher than the respective likelihood for a specific heterogeneous mix of substances. This in the sense that the multiplication of a macro-molecule with a polymeric structure, consisting of x building blocks, only requires conditions that provide molecules of the same type, the respective monomers. In the case the macromolecule has a non-polymeric structure, it would need up to x different building blocks. The higher the number of x, the greater the difference would be.

In summary, *Interaction Theory* allows the following conclusion about the beginning of life: the first phase of evolution must have been dictated by the relatively limited catalytic capabilities of the molecular interaction communities from which the generation sequences of life originated. Thus, the molecular forms were dependent on environmental conditions which had to fulfil highly specific requests and therefore a very low reliability to be available for an ongoing multiplication. This situation can be seen as a decisive weak point for the probability of multiplication of the early generation sequences. To overcome the weak point, the early multiplying forms needed to adapt to progressively more reliable, while less specific and easier available conditions. In this way the probability of multiplication could have increased over time. In the end, a corresponding development would result in the adaptation to regenerating niches, such as those occupied by autotrophic prokaryotes. It can be assumed that this type of niche is very reliably available, because it needs not much more than relative simple carbon and energy sources plus some minerals (e.g. Knoll A.H. 2003). In the end, *Interaction Theory* provides a logical explanation of why the first phase of evolution stretched from simple molecules to solitary multiplying, prokaryotic forms. In contrast herewith, common evolutionary theory has problems to provide a comparable explanation for this phase of evolution (see e.g. Kauffman S. 2003 or Stewart I. 2003).

The law-like process of complexity increase in the first part of evolution

In this section, I conclude the discussion of the first part of evolution with arguments as to why the early generation sequences were subject to the law-like process of complexity increase. For this I want to start with the question: why would key innovations be necessary to strengthen the identified weak point of the generation sequences at the beginning of evolution and what would those look like?

It was said that the proposed molecular interaction communities or chemical networks at the beginning of life are characterized by the capability to facilitate their own multiplication. More concretely this means that the different molecules had to play an essential role in catalysing or facilitating the formation of at least one other type of interaction partner in the network. And if, as proposed, the early interaction partners had a polymeric structure this would mean that they were involved in the connection of the respective monomers of which their interaction partner consisted[7]. In the end this mutual participation in their respective formation can be seen as the basis for the development of the transcription and translation mechanism which is in the centre of the cellular metabolism. Here the three macro-molecules DNA, RNA and protein play an essential mutual role in their formation: RNA catalyses the peptide bond and dictates the amino-acid sequence of the cellular proteins, protein catalyses the bond between the nucleic acid monomers of DNA and RNA, and DNA dictates the sequence (Campbell N.A. et al. 2008).

As said, the first phase would end with the emergence of multiplying forms, such as autotrophic prokaryotes, which are able to grow with not much more than sunlight, CO_2 and some minerals. The metabolism of these forms must therefore be able to synthesise all essential substances, such as amino-acids or nucleotides. The development from relatively simple -since spontaneously emerging - molecular interaction communities, towards correspondingly autotrophic forms requires therefore the acquisition of enormous biochemical capabilities. For this it is necessary that the early molecular interaction communities included more and more members, which were no longer directly involved in the formation of interaction partners, but rather catalysing other chemical reactions. An example would be the

---

[7] This does not exclude the need for external catalysts, such as certain minerals, as proposed by several authors (e.g. Wächtershäuser G. 2000). In this case the role of the interaction partners could have been to influence the precise sequence of the newly formed molecules.

transformation of a molecule into a suitable building block for the multiplication reaction. In the case the corresponding molecule would be easier or more reliably available in the surrounding medium, compared to the building block itself, the new catalytic capability allows the multiplication under less specific conditions. As I will explain below, from the *Interaction Theory* point of view this kind of new catalytic qualities corresponds to key innovations.

The stepwise acquisition of corresponding key innovations, first by the molecular interaction communities and then by the first cells, could continuously increase the probability of multiplication in the first phase of evolution. Along this way, the basic cellular metabolism could have been assembled step by step. Each of the steps would correspond to a round of the law-like process, driven by environmental changes that caused the vanishing of a hitherto essential factor. The consequence would be the disappearance of molecular interaction communities, later cells, which depend on this factor, and the ascent of corresponding forms with a key innovation that is capable of synthesizing the factor from a more reliable precursor. For example, some authors believe that certain substances essential for the early phase of life, such as amino acids, may have arrived via meteorites on the early earth (De Duve C. 1995). It is obvious that this type of resource has a limited reliability, a fact that is in line with above argumentation. In summary it can therefore be concluded that after the first molecular interaction community emerged, the law-like process of complexity increase could have kicked off. It produced with each round a new key innovation, corresponding to a new, additional catalytic capability that allowed the multiplication under less specific conditions.

But why would the new catalytic qualities be necessarily key innovations? As defined, key innovations have the following two characteristics. Firstly they allow the adaptation to new environmental conditions, under which the original form cannot grow. And secondly they can result in a higher probability of multiplication under ordinary conditions. Now, if a molecular interaction community or a primitive cell acquires the capability to use a more reliably available substance for its multiplication, both would be the case. On the one hand the multiplication can now occur under new conditions, not possible without the key innovation. And on the other hand, the individual generation sequence can also have a relatively higher likelihood to continue over time under the original, respectively ordinary conditions. The reason is that thanks to the key innovation the multiplication becomes less sensitive with regard to fluctuations of the original substrate.

In addition, key innovations need a dual selection to emerge. Also in this regard it can be concluded that the emergence of corresponding new catalytic qualities would require a dual selection. The reason is that an exclusive selection on fitness would result in a general growth contest. For molecular interaction communities this means that all molecules become multiplied, and over time those multiplying the fastest prevail. As outlined in the first part, such a contest will inevitably run into a thermodynamic equilibrium situation that stops the generation sequences. Thus, an exclusive selection on fitness would not be possible in the first phase of evolution. In addition, the already quoted Spiegelman trial (Spiegelman S. 1970) illustrates nicely the structural consequence of a growth contest for molecular forms. The ongoing selection on a fast multiplication caused a strong simplification of the respective molecules, here RNA. As a consequence it is difficult to imagine how the sole selection on fast multiplication, thus an exclusive selection on fitness, would be able to cause the acquisition of new, structurally more sophisticated qualities (compare Kaufmann S. 2003); in particular, if this could not happen in one accidental variation step. Thus, it makes much more sense to assume that the acquisition of new, structurally sophisticated qualities would need some kind of investment. And this would require dual selection or symbiosis, as outlined previously.

An example of what the dual selection within a molecular interaction community would look like is the following. Let's imagine a simple molecular network in which protein catalyses the linkage between nucleic acid monomers. To do so the protein molecules need a special catalytic site that is given by their specific molecular structure. In addition, however, the molecules need to interact with their respective interaction partners (compare illustration 1). This means that via non-covalent interactions they need to bind to other molecules of the interaction community. Both qualities, the catalysis of a specific reaction and the specific binding to an interaction partner, would be subject to different kinds of selection. In the first case it would be a selection favouring variants with a better catalytic ability, in the second case it would be a mutual selection favouring variants with better interaction qualities. Obviously this corresponds to a dual selection, as it is required for the emergence of key innovations. In the case of the protein example, both the catalytic and the interaction quality are the result of the respective amino-acid sequence and their three-dimensional structure. This means that if a protein variant with a different amino-acid sequence emerges that binds better to their interaction partner(s), it can also have different catalytic qualities. Concrete examples for the emergence of new catalytic features by modifying the binding site to other molecules are catalytic antibodies (see e.g. Wentworth P. Jr, Janda K.D. 2001). In a comparable way, the dual selection within molecular interaction communities could have caused the emergence of new catalytic qualities.

In summary, there are good arguments that the first phase of evolution followed the law-like process of complexity increase, as predicted by *Interaction Theory*. With each round of the process, the probability of multiplication increased and transformed the early molecular forms into prokaryotic forms, able to occupy re-generating niches.

The genetic code

All life on earth is based on the same genetic code. For this reason it was con-cluded that it must have developed in the last common ancestor of all living forms. How this could have happened is not understood (see e.g. Maynard Smith J. and Szathmáry E. 1995). Nevertheless, from the *Interaction Theory* point of view it is possible to link the emergence of the genetic code with what was said about the first part of evolution. A chemical network in which all members are essentially involved in their mutual formation has in principle no need for a genetic code. All members ensure their own formation, in the sense that they help to multiply other members of the respective community, which in turn ensures their own multipli-cation. At the moment however when the network depends on members with other duties, such as transforming precursors into suitable building blocks, the situation changes. Now a solution is required that allows also the formation of those mole-cules, which are no longer directly involved in the mutual formation reaction.

From this point of view, the genetic code looks like the perfect solution as to how interaction communities consisting of DNA, RNA and protein can multiply all sorts of proteins, irrespective of whether or not they are directly involved in the mutual formation. In this context it is also obvious that the emergence of some-thing as sophisticated as the genetic code depends on a strong and constantly given benefit gradient, so that it can develop step by step. The above scenario of a step-wise adaptation to increasingly reliable conditions indicates an answer as to what a corresponding benefit gradient could have looked like. This means that the se-quence of countless rounds of the law-like process of complexity increase would drive the emergence of the genetic code. For the development from highly specific environmental conditions to regenerating niches, the interaction communities had no other choice than to follow the way that caused the development of the genetic code. In the end this means that without the emergence of the genetic code the stepwise adaptation to environmental conditions with an increasing reliability to sustain the multiplication over time would not have been possible. Thus, the ge-netic code corresponds to a key innovation.

Before I continue with other phases of evolution, I would like to make a remark about the end of the first phase. At the moment prokaryotic forms arrived on the scene with their ability to occupy regenerating niches, the overall situation would fundamentally change. Prokaryotes correspond to what I have called biological multiplication bombs. They can grow on nearly all sorts of substrates and with their endospores they can easily spread around the globe. The consequences are therefore irreversible. In a world with prokaryotes, the undisturbed emergence and ongoing multiplication of molecular interaction communities would no longer be possible.

2.3. The emergence of eukaryotes

After the generation sequences have arrived on their journey through deep time in the world of solitarily multiplying prokaryotes, what can be said about the further step of biological complexity increase from the *Interaction Theory* point of view? Again, the basic assumption is that complexity increase goes towards multiplying forms with a relatively higher probability of multiplication, so that the individual generation sequence has an on average higher likelihood to continue over time.

With prokaryotes capable of multiplying under conditions which can regenerate after their exhaustion by exponential growth, the probability of multiplication cannot continue to rise in the same way as during the first phase. For prokaryotes the probability of multiplication is no longer dictated by a low reliability of the environmental conditions necessary for their multiplication. It is rather dictated by their growth behaviour, oscillating between exponential growth and stagnation. As I will suggest in the following, for this kind of multiplying form there exists therefore only one way how the probability of multiplication can continue to rise. And this way will eventually lead the generation sequences into irreversibly exhaustible or destroyable niches and thus towards the evolution of more complex organisms with sexual reproduction. The transition between the two fundamental types of niche is apparently connected with a main step of biological complexity increase, the emergence of eukaryotic cells from structurally simpler prokaryotic ancestors. In the following I will provide arguments as to why the principles of *Interaction Theory* are in line with what is known about the evolutionary history of eukaryotic cells.

The weak point of the probability of multiplication for prokaryotes

In what way can prokaryotic forms start to increase the probability of multiplication of their generation sequences? Single cell organisms multiply by cell division, and the time between two consecutive divisions impacts therefore the possible growth rate. This has the consequence that the generation sequences of bigger cells require a relatively higher probability of multiplication compared to smaller ones. And this in turn allows the assumption that a significant rise in the probability of multiplication for unicellular forms goes along with a substantial increase in cell size. And indeed this is exactly what has happened during evolution: relatively small-sized prokaryotes developed into much bigger unicellular eukaryotes (Knoll A.H. 2003).

In the world of regenerating niches occupied by solitary multiplying forms such as prokaryotes, the probability of multiplication is dictated by the rollercoaster-like growth behaviour, oscillating between exponential growth and stagnation. In this regard, the niche's capability to tolerate exponential growth is crucial. And obviously, niche factors exhaustible by exponential growth are important in this context. For example, bacteria such as *E. coli* need for their growth an external carbohydrate source (see e.g. White D. 2007). This source becomes inevitably exhausted by the exponential bacterial growth and needs therefore the ability to regenerate over time. In this regard it is evident that not all potential resources can fulfil this demand. For this reason it can be expected that the number of regenerating niches should be rather limited than endless. In contrast herewith it will turn out that a comparable limitation does not exist for irreversibly exhaustible niches, occupied by reproductively interacting forms, such as sexually reproducing species (later I will come back to this point). This difference can explain why the observable number of forms, respectively species, as an indicator for the number of existing niches, seems to be lower for prokaryotes than for eukaryotes (see e.g. Knoll A.H. 2003 or Mora C. et al. 2011).

Already mentioned examples for resources which can regenerate after exhaustion by exponential growth, are relatively simple substances, such as carbon dioxide or methane. On the other hand, other organisms can also serve as essential resources for growth. The first would correspond to what is called autotrophy and the second to heterotrophy. In the case other organisms serve directly as a decisive resource, as in a prey-hunter or host-parasite relationship, this can be described as a negative coupling of the multiplication between different types of multiplying forms. Negative means that the growth of one party, the hunter or the parasite, burdens on the population of the other, the prey or host. As already mentioned in the first part, relationships of this kind can only be sustainable if the prey or host does not become endangered by the exponential growth of the hunter or parasite. And precisely here lies a decisive weakness of prokaryotes: the negative coupling of the multiplication between prokaryotic forms bears the risk of provoking negative, irreversible changes, which are comparable to an irreversible exhaustion of resources. As a consequence it may reduce the prospect of the respective generation sequences of continuing over time.

To make this more concrete, let's consider a prey-hunter situation between prokaryotic forms, in which the prey is essential for the hunter. The multiplication of biological forms goes along with the emergence of new variations. This means

118

that the hunter and the prey population are not absolutely homogeneous, since they show genetic differences between the individuals (see e.g. Jurkevitch E. 2007). In this regard it is obvious that not all hunter variants can grow equally on all possible prey variants. And precisely this can be seen as the reason why prey-hunter or host-parasite relations between prokaryotes would face restrictions. For example, selection on fitness will favour those hunter-variants which grow successfully on the most frequent prey-variants. At the same time, newly emerging prey variants will have a decisive advantage, if they are protected against the most frequent hunter variants. If this is the case, these new prey variants can prevail over time. As a consequence, the prokaryotic hunter may even risk losing its prey, if not the changing situation would provoke a counter-selection that favours hunter variants, able to grow successfully on the new type of prey. In conclusion this means that the negative coupling of the multiplication between prokaryotic forms can only be sustainable in the long term if none of the involved parties loses the corresponding arms race. From the *Interaction Theory* point of view this inevitable arms race, with its potential negative consequences, corresponds to a decisive weak point for the generation sequences of prokaryotes.

As mentioned previously, the above situation is often described in the literature as Red-Queen hypothesis. Decisive for the sustainability of this type of relationship must be the frequency by which new variants appear on both sides. For solitary multiplying forms this frequency depends in general on the number of mutations occurring per time. The mutation rate per reproduction or lifecycle is, however, limited; otherwise multiplying forms cannot survive in the long-term (Eigen M. 1971). For a hunter-prey or host-parasite couple this means that both parties can in principle produce a comparable number of mutations per lifecycle. But this would also mean that the possible number of new variants per time depends on the respective generation time. As a result it can therefore be concluded that the difference in generation time between prokaryotic hunter and prey should not be too big. The reason would be that a hunter with a much longer lifecycle, compared to its prey, would be in danger of not producing enough variation per time. Consequently it may lose the arms race with the faster-multiplying prey. In summary, the negative coupling of the multiplication between prokaryotic forms should only be sustainable in the longer term, if the difference in the generation times between the parties is not too big. This would obviously make it difficult for prokaryotic hunter cells to become much bigger than their prey, even if it would be beneficial for other reasons.

The above considerations are in line with the previously already mentioned observation that bacterial predators follow the rule "small feeds on big" (Jurkevitch E. 2007). This supports the conclusion drawn that the sustainable, negative coupling of the multiplication between prokaryotes faces in principle restrictions. In contrast, corresponding restrictions do not seem to exist for eukaryotic microorganisms, which feed happily on much smaller and faster-multiplying bacteria (for protists see e.g. Campbell N.A. et al. 2008). The fact that eukaryotic bacteria hunters are much bigger than their prey can therefore be seen as the visible sign that eukaryotic cells did overcome the identified weak point of the generation sequences of prokaryotes.

In summary, two points can be seen as relevant for the further complexity increase, after prokaryotic forms evolved. Firstly, for unicellular microorganisms multiplying via cell division, a rising probability of multiplication is linked to an increasing cell size. Secondly, the negative coupling of the multiplication between prokaryotes faces restrictions, in order to be sustainable in the long-term. This concerns great differences in size between prokaryotic prey and hunter or host and parasite. From the *Interaction Theory* point of view, this corresponds to a principal weak point that burdens on the probability of multiplication of prokaryotic generation sequences exposed to a negative coupling of their multiplication. It can therefore be predicted that the law-like process of complexity increase produces key innovations that strengthen this specific weak point. In the end it means that eukaryotic microorganisms can feed on much smaller bacteria, because they own a respective key innovation – see below.

Before closing this section I want to come back to the host-parasite relation between bacteriophages and their prokaryotic hosts, which was already discussed in the first part. There it was mentioned that the smaller and therefore faster-multiplying bacteriophages are potentially at risk of eradicating their larger and slower-growing host. And it was also said that the characteristics of virulent and temperate phages can be seen as visible consequence of this risk. Temperate phages insert for example their DNA into the genome of the host and cause resistance against further infections (Clokie M. R.J. et al. 2011). And the virulent phages, such as T4, which kill their host by lysis after infection, correspond to the most complex viral structures in nature (Yap M.L. and Rossmann M.G. 2014). This would reduce the complexity gap between the bacteriophages and their hosts, in order to allow a sustainable host-parasite relationship.

The observation that bacteriophages are very frequent in nature (Clokie M. R.J. et al. 2011) may also explain why prokaryotes remained tiny during evolution, irrespective of whether they are auto- or heterotrophic. The explanation would be that bacteriophages, such as T4, have reached an upper physical limit regarding the maximum possible complexity of corresponding self-assembling structures. If this is the case, then prokaryotes had to remain tiny, because otherwise they would fall victim to these widespread parasites.

The significance of an increasing amount of DNA

Big eukaryotic single-cell organisms, such as protists, feed on much faster-multiplying bacteria (see e.g. Bonner J.T. 2006). How can this be explained, in the light of the previous reflections? The suggested answer is closely linked to a main characteristic of eukaryotic cells: a significant higher amount of DNA per cell, compared to prokaryotes.

From the *Interaction Theory* point of view, the benefit of a significantly higher amount of DNA per cell would be relevant for the generation sequence and not for the individual form. The reason is that it allows the conservation of genes which may not be essential for the multiplication success in the here and now, but can potentially become important several generations later. A good example is the immune system, which provides a preventive defence against a high number of different antigens with which it may never come into contact (e.g. Campbell N.A. et al. 2008). In other words, the immune system is based on the principle of a preventive conservation of qualities which may or may not be beneficial in a particular generation. This principle can be seen as a solution for the prey-hunter relationship discussed: the prokaryotic hunter cell needs to recognize, kill and digest its prey. To succeed with all possible prey variants equally, the hunter would obviously need more qualities than for one single variant only. As an example, the recognition of the prey cell may need one specific receptor per variant. Because qualities such as specific receptor proteins need to be reflected in the genome, the capability to grow on many different variants would need relatively more genes. Thus, the capacity to grow on all possible prey variants would demand more DNA and thus a larger genome. On the other hand, prokaryotes keep their genome tight. Genes with little contribution to fitness are more susceptible to deletional mutations (see e.g. Ochman H., Davalos L.M. 2006). A main reason can be seen in the fact that the prokaryotic DNA replication starts from one replication point only, and the replication time of the whole genome influences therefore the time between two cell divisions. Consequently, the variants with less DNA multiply relatively faster and can prevail, as long as they own all qualities necessary in the here and now. This means that selection on fitness rewards the recognition, binding and digestion of the predominant prey variants. However, it does not help much to conserve qualities which are not relevant for the fitness of the particular form, even if one of these qualities may be of advantage in a future generation with regard to new prey variants. This inability for a preventive conservation of genetic information can be seen as a reason why the negative coupling of the multiplication between prokaryotes causes the red-queen effect.

To change the situation, the cells would have to avoid or at least to significantly slow down the loss of genetic information without a clear fitness benefit in the here and now. For this, it would be necessary to organize the cell division in a new way and to overcome the central role of the DNA replication as timer for the duration of the cell reproduction. This can obviously be done by starting the DNA replication from more than one point only, as is the case with eukaryotes (see e.g. Maynard Smith J. and Szathmáry E. 1995). Consequently the DNA length becomes less important, which corresponds to a precondition for the conservation of genetic information with no immediate fitness benefit. Thus, for a bacteria hunter this would be a necessary step in order to escape the red-queen effect and to conserve qualities that might be relevant for new prey variants. This in turn would allow an increasing cell size and generation time, without provoking the risk of losing the arms race against a much faster-multiplying prey.

In summary, the insight that complexity increase goes into the direction of a rising probability of multiplication allows conclusions about the course of evolution, once the state of prokaryotes is reached. A rise in the probability of multiplication for the generation sequences of single cell organisms demands that the respective cells become significantly bigger and their generation time thus much longer. Here however, the negative coupling of the multiplication between prokaryotes, as well as between prokaryotes and bacteriophages, represents an issue because of the red-queen effect. The latter is responsible for the fact that a much slower multiplying party is at risk of losing the respective arms race. This would act against strong differences in generation time, and therefore complexity, within prey-hunter and host-parasite relations. However, the further complexity increase towards big cells in a world of tiny prokaryotes would in principle be possible with cells that are characterized by a preventive conservation of qualities. This means that these forms would compensate their lower number of variations per time by conserving genetic information much better than is typical for prokaryotes. This makes it possible to keep genes longer, even if they do not have a fitness benefit in the here and now, but may potentially become important some generations later. The visible sign of complexity increase within a world of prokaryotes would therefore be the emergence of big single cell bacteria hunters with a significant longer generation time and a much larger amount of DNA.

The above considerations evidently match the general key differences between eukaryotes and prokaryotes (for a summary of the differences see e.g. Margulis L. 1993). Eukaryotic single cell organisms are much bigger, have longer generation times and significantly more DNA per cell than prokaryotes. In addition, eukaryotes have organized their genomes differently. It is no longer one circular molecule per cell, but different, linear chromosomes, of which each has multiple starting points for the DNA replication. Hence, eukaryotes have therefore surmounted the identified weak point of prokaryotes regarding the negative coupling of the multiplication. In addition, the above considerations also supported by the fact that the known, most primitive eukaryotic single cell organisms, such as Giardia (see below), are indeed feeding on bacteria. In the end this shows again that *Interaction Theory* allows predictions about the course of evolution which are in accordance with reality.

The argument that a preventive conservation of genetic material is important for microorganisms is also supported by the fact that the so-called conjugation is widespread with prokaryotes as well as eukaryotic ciliates. Conjugation describes here the process in which genetic material, mostly in the form of so-called plasmids, is transferred from a donor to a recipient cell. It corresponds therefore to a particular kind of interaction between different types of cells, the donor and the recipient. And most important, it is well known that conjugation offers the benefit of keeping qualities such as antibiotic resistance in a common gene pool. This reduces the risk of losing the kind of qualities which may not provide a benefit in the here and now, but can be essential some generations later (for plasmids see e.g. Bennett P. M. 2008).

Complexity increase via symbiosis - the *Endosymbiotic Theory*

In the following let's turn to the question of how the emergence of eukaryotic cells in ecosystems dominated by prokaryotes could have occurred from the *Interaction Theory* point of view.

A key insight of *Interaction Theory* is that solitary multiplying forms need symbiotic interactions to acquire key innovations necessary for complexity increase. Following on from what was said in the previous section, the key innovation would correspond in this case to the capability to conserve significantly more DNA per cell. For this the cells would need a new and different organisation of the reproduction cycle, starting the DNA replication from more than one point, as is the case for eukaryotes. Thus, the emergence of cells which are characterized by a respective key innovation should be the result of a symbiosis between different prokaryotic cells. And indeed, this assumption is in accordance with the well-established *Endosymbiotic Theory,* seeing symbiotic events at the origin of eukaryotes.

That the evolution of eukaryotes is the result of symbiotic events is strongly advocated by Lynn Margulis and others (see Margulis L. 1993). The corresponding theory states that eukaryotic cells are the product of different symbiosis events. In a first step Prokaryotic cells of various types fused with each other and the resulting hybrid developed into the first nucleated cell. The above-mentioned key innovation, to switch the DNA replication per cell from one single starting point to

123

multiple starting points together with the capability to preserve more DNA per cell, should have occurred during this step. In an independent, following step the nucleated cells would then have integrated an additional prokaryote, very probably belonging to the α-proteobacteria, which became the mitochondria. And in a third step, nucleated cells with mitochondria integrated one more prokaryote. This time it was a photosynthetic form belonging to the cyanobacteria which became the chloroplast in plants and algae. Today *Endosymbiotic Theory* is widely accepted as explanation for the emergence of eukaryotes, even though the details are still unclear (see e.g. Embley T. M. and Martin W. 2006, regarding the open questions about the nature of the host that acquired the mitochondrion).

In this regard it is interesting that the eukaryotic genome shows traces of different types of prokaryotes. So it was suggested that at least one archaebacterium was involved in the formation of eukaryotes. In addition, thousands of eukaryotic operational genes are seen as eubacterial (see Martin W. and Müller M. 1998). Obviously this favours speculation that the decisive symbiosis event at the birth of eukaryotes involved more than two partners only (Margulis L. 1998). And irrespective of the exact nature of the symbiosis partners, from the *Interaction Theory* point of view it supports also the assumption that it was necessary to organize in a new way the DNA replication in the resulting bigger entity - this in the sense that a coordinated DNA replication of the different partners should have been essential for a successful cell division of the new hybrid cell. And this could have eventually resulted in a cell which started its DNA replication cycle from more than one point, and conserved bigger genomes with qualities not relevant for the multiplication success in the actual generation.

On the other hand, would it also be thinkable that the eukaryotic cell emerged from a prokaryotic ancestor without any symbiotic unification? From the *Interaction Theory* point of view the answer is clearly no, for the following reasons: with their exceptional biochemical power, prokaryotes are able to grow on all possible substrates and as solitary multiplying forms they experience a strong selection on fast replication. This makes it difficult to imagine how and why these multiplication bombs would have started a gradual, solely selection on fitness-driven development towards slower multiplying forms with a much higher DNA amount per cell. In addition, the clustered organization of living forms in prokaryotes and eukaryotes indicates that the newly emerging eukaryotic cell would have adapted to a new kind of niche. As outlined in the first part, the emergence of a new phylogenetic cluster within established ecosystems can be seen as the sign that a dually selected key innovation made it possible to occupy a different kind of niche, not accessible to the other forms at that time. As further outlined previously, if this were not be the case, it should be expected that eukaryotic cells emerged gradually and in several groups of prokaryotes. This, however, is clearly not the case (see e.g. Maynard Smith J. and Szathmáry E. 1995).

The *Endosymbiotic Theory* is also supported by observations that prokaryotes are frequently engaged in symbiotic relationships (Margulis L. 1998). With a high prevalence of symbiosis it may therefore only be a matter of time until a new type of cell emerges via symbiotic unification which is able to occupy a new niche not accessible to prokaryotes. Micro-paleontological findings suggest that the first prokaryotic-like cells appeared surprisingly early in evolution, about 3.6 billion years ago (Knoll A.H. 2003). Compared with this it seems that the breakthrough of eukaryotes occurred much later, very probably it took more than another 1 billion years (again Knoll A.H. 2003). The reason for this very long time may be simply the fact that both cell types differ in more than their relative amount of DNA and their organisation of the DNA replication. Eukaryotes own, for example, a specific cytoskeleton and the capability for phagocytosis, both not known in prokaryotes. Respective qualities are obviously very useful for cells which need to catch and digest smaller sized prey – see following. Consequently eukaryotes and prokaryotes differ in several aspects. Therefore, a sequential acquisition of several key innovations may have been necessary in order to produce the kind of eukaryotes as we know them. This could explain why it took so long.

The emergence of eukaryotes via the law-like process of complexity increase

Interaction Theory predicts that the complexity increase of solitary multiplying forms requires the emergence of a key innovation via symbiosis, what is in accordance with the *Endosymbiotic Theory*. In the following I want now to discuss whether the facts that are known about the emergence of the first eukaryotes are also in accordance with the principle of the law-like process of complexity increase.

As a consequence of the identified weak point for the generation sequences of prokaryotic hunters, it follows that the next step of complexity increase would concern forms feeding on prokaryotes. In this regard, a bigger cell size can represent an advantage for unicellular organisms feeding on smaller prey. This would in particular be the case if the hunter cell engulfs the prey for digestion. Consequently it can be assumed that in a world of tiny prokaryotes, big hunter cells with the ability for phagocytosis can access a new kind of niche. For this, however, they need the discussed key innovation that allows a preventive conservation of genetic information. As said, the generation sequences of big hunter cells without this key innovation would not be sustainable in the long term, because of the risk of losing the unavoidable arms race with the much smaller prokaryotic prey. And indeed, this scenario is in accordance with the fact that primitive eukaryotic cells, such as the anaerobic single cell organism *Giardia*, feed on prokaryotes, by engulfing their smaller prey (for *Giardia* see e.g. De Duve C. 1995). Thus, this supports the view that the first eukaryotes were indeed corresponding hunter cells, as predicted by *Interaction Theory*.

While *Giardia* lives under anaerobic conditions, most eukaryotic single cell organisms such as protists, which feed on bacteria too, are aerobic (Campbell N.A. et al. 2008). In contrast to *Giardia,* these cells own mitochondria, to use oxygen. It is generally believed that the acquisition of mitochondria by early eukaryotes would have enabled the change from an anaerobic to an aerobic lifestyle (see e.g. Maynard Smith J. and Szathmáry E. 1995). This means that thanks to the acquisition of mitochondria, the big hunter cells were able to continue their generation sequences also in oxygen containing atmosphere.

In the case of chloroplasts, the situation may have been slightly different. Eukaryotic forms with chloroplasts, such as unicellular algae, are able to grow with carbon dioxide and sunlight as main resources. In this respect they are not different from cyanobacteria. On the other hand, also in this case a decisive difference is the much bigger cell size of the eukaryotic forms. Again it can therefore be assumed that thanks to their size the first photosynthetic eukaryotes were able to occupy a new kind of niche. Corresponding environmental situations in which size matters are obviously not appropriate for tiny cyanobacteria. This conclusion is in line with the fact that single cell algae are among the largest unicellular organisms known (Campbell N.A. et al. 2008).

The evolution of eukaryotic cells is not well documented in the fossil record and lies therefore to a large extent in the dark (Knoll A.H. 2003). On the other hand, it is believed that the breakthrough of nucleated cells with mitochondria is a result of the global rise of atmospheric oxygen levels about two billion years ago (Embley T. M. and Martin W. 2006). Obviously the rising oxygen represented a global environmental catastrophe for the then dominant anaerobic forms. In accordance with the law-like process of complexity increase, it destroyed certainly the ecosystems thus far established, and opened the way for a new eukaryotic diversity.

In conclusion, the relatively few facts that are known about the early evolution of eukaryotes support the view of *Interaction Theory* that the emergence of eukaryotic cells followed the law-like process of complexity increase.

2.4. The emergence of sexual reproduction

Thus far I have followed the journey of generation sequences from spontaneously emerging molecular interaction communities to unicellular eukaryotes. This resulted in the conclusion that during this journey, the probability of multiplication has progressively increased and with large eukaryotic cells eventually reached its first peak. In the following I will now discuss how generation sequences can continue this development after the stage of eukaryotic cells with mitochondria and chloroplasts has been reached. As the reader will see, this leads directly to sexually reproducing species.

The probability of multiplication of unicellular eukaryotes

According to *Interaction Theory*, the complexity increase during evolution is the result of a law-like process that produces in ever-repeating rounds multiplying forms of which the individual generation sequence shows a progressively higher likelihood to continue over time. Following this logic, what is the next step after ecosystems with eukaryotic micro-organisms have been reached?

Unicellular eukaryotes without any form of reproductive interaction correspond to solitary multiplying forms. Hence, their probability of multiplication is dominated by the continuous change between phases of exponential growth and stagnation. In the period of growth the cells multiply and the number of generation sequences strongly increases; in the following period of stagnation, most of the generation sequences end again. For solitary multiplying forms, trapped in a general growth contest, this permanent rollercoaster is unavoidable in an environment where endless growth is not possible. The difference between prokaryotes and eukaryotes in this regard is that the larger and more complex cell structure of eukaryotic micro-organisms results in a significant longer generation time and therefore a slower growth under favourable conditions (see e.g. Bonner J.T. 2006). Consequently their probability of multiplication needs to be on average higher, compared to prokaryotes.

What kind of theoretical solutions exist for unicellular organisms, to further increase the probability of multiplication? An ongoing increase in cell size could be in principle a possibility. In other words, in case eukaryotic cells continue to become bigger and their generation time becomes ever longer, this requires a further rising probability of multiplication in order to be sustainable. But I will exclude this development. The reason is that it is a generally accepted assumption that the size of unicellular organisms is restricted, because cellular processes occur on the microscopic level and cannot be stretched out over a macroscopic dimension (De

Duve C. 1995 or Bonner J.T. 2006). However, is another solution thinkable? The answer is yes, if it were possible to overcome the roller coaster-like growth pattern that burdens on the probability of multiplication for solitary multiplying forms. Thus the up-and-down of growth and stagnation can be seen as a weak point for the corresponding generation sequences. Its strengthening would require multiplying forms with a more even growth pattern, which avoids the dramatic changes in the number of generation sequences.

How can multiplying forms achieve a more even growth pattern that allows a further increasing probability of multiplication? The answer was given in the first part: it requires a competition for multiplication success, which can occur without a general growth contest. As outlined, this would be possible with reproductive interactions such as sexual reproduction. The associated mutual selection within the respective interaction community is the key for a more even growth pattern. As a result, the sustainable occupation of the second fundamental type of niche becomes possible. The latter are environmental conditions with regenerating resources and a sufficient probability of being present over time, which however do not tolerate an exhaustion or destruction by exponential growth. Consequently they cannot be occupied in a sustainable way by solitary multiplying forms.

For the next step of complexity increase it can therefore be predicted that for a further substantial increase in the probability of multiplication, reproductively interacting forms would be necessary, showing a mutual selection on interaction competence. And indeed, the emergence of sexual reproduction was the next major step in the evolution of eukaryotes (see e.g. Knoll A.H. 2003). Again, the conclusions derived from *Interaction Theory* are in accordance with what has occurred during evolution.

In summary, after solitary multiplying forms, first prokaryotic and then eukaryotic, emerged in exponential growth tolerating niches, the further rise in the probability of multiplication would need reproductive interactions, such as sexual reproduction. As a consequence, the corresponding generation sequences can now continue their journey in niches which can become irreversibly exhausted or destroyed by exponential growth. And this in turn means that the qualities associated with sexual reproduction, such as genetic recombination, correspond to a key innovation in the sense of *Interaction Theory*.

Genetic recombination and niche adaptation

The emergence of sexual reproduction needs the genetic fixation of the corresponding qualities, such as those responsible for the genetic recombination. As discussed in the first part however, genetic recombination may be good for the generation sequences, but under regular niche conditions it does not normally increase the multiplication success of the individual form in the here and now. As said, from the *Interaction Theory* point of view, this would be true for all key innovations, and their fixation requires therefore the law-like process of complexity increase. In the following I want to discuss genetic recombination under this aspect.

The recombination of genetic information is an essential part of sexual reproduction that occurs during meiosis and the production of gametes. Basically it means that homologues chromosomes from both parents exchange DNA and can therefore produce descendants with a modified genetic profile (Maynard Smith J. and Szathmáry E. 1995). In the first part, genetic recombination in sexually reproducing forms was discussed in the context of a sustainable adaptation to niches irreversibly exhaustible or destroyable by exponential growth. It can separate genes causing a high interaction competence, or sexual attractiveness respectively, from those responsible for a high multiplication potential. Hence, genetic recombination can build a barrier against a development towards a progressively higher multiplication potential, which can eventually cause unsustainable growth; this by ensuring that mutual selection and selection on fitness may over the generations not always point in the same direction. In the following I will now discuss why genetic recombination could have emerged in eukaryotic micro-organisms.

In the previous chapter it was deduced that the first eukaryotes occupied environmental conditions as niches which do not tolerate the fast specialization as is known from prokaryotes. Fast specialization means here that the cells lose relatively easily qualities with little or no contribution to fitness (see e.g. Bennett P. M. 2008 or Cooper S. 2012). In comparison, eukaryotes do not keep their genome in the same way compact as prokaryotes, and thus show a significant higher amount of DNA per cell. As mentioned, a corresponding key innovation in this regard would be a different coordination between cell division and DNA replication, compared to prokaryotes. For this reason it was concluded that eukaryotes can better conserve genetic information with no contribution for the individual fitness in the here and now. This was described as the capability of a preventive conservation of genetic information, which allows not only conserving qualities that are essential at the moment, but also those becoming several generations later potentially essential. With this it would be possible to occupy in a sustainable way niche situations not possible for prokaryotes, as in the previously given example of big cells feeding on much smaller bacterial prey. In the end preventive conservation of genetic information means that eukaryotic micro-organisms can keep their genetic profile broader than prokaryotes. The subsequent emergence of genetic recombination would fit into this scheme. It corresponds to an additional step in how to act against a specialization of the genetic profile. While selection on fitness favours a maximisation of the multiplication under the given conditions, the recombination of the genetic information can break up the combination of genes, irrespective of whether it is favourable or not under the given situation. From the *Interaction Theory* point of view, recombination may therefore have emerged in eukaryotic micro-organisms, because it can keep the genetic profile broader, compared to a specialisation on prevailing conditions. Here it should be mentioned that the emergence of diploid cells can also be seen as part of this scheme, but I do not want to discuss this further.

This role of genetic recombination is also suggested by conventional evolutionary theory and confirmed by the behaviour of unicellular organisms with periodic sexual reproduction. While being exposed to favourable environmental conditions they multiply as fast as possible via mitosis. The moment lesser favourable conditions appear, they switch to meiosis and the formation of gametes. The recombination during meiosis causes new allele combinations which may help to deal with the less favourable conditions (Stearns S.C. and Hoekstra R.F. 2005).

On the other hand and in contrast to conventional evolutionary thinking, genetic recombination would not only have the benefit of facilitating the adaptation to new environmental conditions. For example, algae species of the same genus reproduce in freshwater mostly mitotic, while in the ocean mostly sexual. If recombination during meiosis only had the purpose of facilitating the adaptation to new environmental conditions, one might wonder why it is not the other way around (see Wesson R. 1991). The ocean is obviously the habitat much less affected by change, compared to freshwater. Therefore it should be expected that freshwater species are more in need of genetic recombination helping them to cope with the more frequent environmental challenges (see again Wesson R. 1991). However, from the *Interaction Theory* point of view with regard to the role of genetic recombination, this observation makes sense. In habitats with frequent environmental changes, multiplying forms should be less in danger of specialising too strongly in particular conditions. From one generation to the next the forms can face a changing situation. As a consequence, in this situation selection on fitness should favour the conservation of a broad, less specific genetic profile. In comparison, under very stable environmental conditions, selection on fitness should favour individuals that are highly adapted to this stable situation. This means that over the generations the genetic profile risks progressively specialising in the prevailing environmental conditions. This however, may become a dead end for the respective generation sequences in the case the habitat suddenly changes. With a specialized genetic profile the forms may no longer own the necessary qualities to deal with the new situation and become extinct. Under corresponding niche situations, a selection on fitness driven specialisation would therefore represent a weak point for the probability of multiplication of the respective generation sequences. In the end this means that the relative frequency of environmental changes dictates the behaviour of the algae in oceanic and freshwater habitats: in the ocean the species would be more in need of genetic recombination, and therefore sexual reproduction. It acts against the risk of an ongoing specialisation of their genetic profile, provoked by the relatively steady environmental situation. In the freshwater however, the relatively frequent environmental changes keep their genetic profile sufficiently broad.

In summary, the recombination of the genetic material during meiosis can be seen as a further step of unicellular eukaryotes in overcoming a weak point for their probability of multiplication under certain niche situations. It reduces the risk of a specialisation of the genetic profile on environmental conditions that are stable over many generations, but yet can change suddenly. In terms of *Interaction Theory*, genetic recombination would therefore correspond to a key innovation that strengthens this weak point and allows the respective forms to adapt to a new type of niche that they could otherwise not occupy in a sustainable way.

Interesting in this regard is that the emergence of genetic recombination can also be seen as the result of a unification of cells, as in the case of the emergence of the eukaryotic cell itself. This time however, the unification of two cells, the gametes, is followed by meiosis and therefore only temporary – see below. Furthermore, it makes sense to assume that genetic recombination also helped the corresponding generation sequences to better survive a catastrophic event or change that destroyed established ecosystems. Its effect of keeping the genetic profile broad would provide under the completely new conditions a benefit, compared to the more specialised asexual forms. This corresponds exactly to what can be expected from the *Interaction Theory* point of view.

The emergence of sexual reproduction in unicellular eukaryotes

The course of a sexual reproduction cycle repeats the same principle as was previously discussed in the emergence of eukaryotic cells. That means the unification of different cells causes a new type of cell, able to occupy a new kind of niche. In this case, the unification means the fusion of two gametes into a diploid zygote. Because the subsequent meiosis divides the diploid cell again into gametes, the unification is here only temporary. The emergence of this process during evolution could have been driven by the benefits of genetic recombination, as detailed in the last section. Sexual reproduction does, however, not only mean the recombination of genetic information. From the *Interaction Theory* point of view, it corresponds also to an entry ticket into niches which are irreversibly exhaustible by exponential growth. As said, this type of niche can only be occupied by interaction communities with mutual selection on interaction competence. Now, what would be the reason why such interaction communities emerge?

The fusion of two haploid gametes to form a diploid zygote is a key event during the sexual reproduction cycle (see e.g. Halliday T.R. 1994). And here can be the answer to above question: the first interaction community with mutual selection would have been the result of the probably inevitable development from the so-called isogamy towards anisogamy. While isogamy means the possession of iso-gaments of a single kind, anisogamy means the differentiation of the gametes in a big egg cell on the one hand and smaller, very mobile sperm cells on the other. In between lies the situation in which the gametes belong to two different mating types (Maynard Smith J. and Szathmáry E. 1995).

It is generally assumed that the original condition was isogamy, thus all gametes being of the same kind (Halliday T.R. 1994). As described above, corresponding sexually reproducing single cell organisms with isogametes would have emerged, because genetic recombination allows the sustainable occupation of a new type of niche. But as long as all gametes belong to the same type and can therefore fuse with each other more or less unrestricted, these communities lack a real mutual selection on interaction competence. The situation changes however, if the gametes become split into two different mating types. Now, a zygote can only be formed by the fusion of gametes belonging to different types. As a consequence, sexual reproduction becomes from now on a real reproductive interaction, in which two parties require each other for their multiplication. This mutual dependence will obviously provoke a competition for mating partners.

How can the transformation of isogametes into different mating types occur? A plausible explanation builds on the assumption that a mixture of cell organelles from both gametes may be a disadvantage for the zygote (see e.g. Maynard Smith J. and Szathmáry E. 1995). The solution would be the split of the gametes into donors and acceptors of organelles. With this distinction, the development into two different mating types occurs if the donors begin to fuse only with the acceptors and vice versa. The transformation can be driven by selection on fitness, because of the benefit for the zygote of receiving only organelles from one gamete (see Hurst L.D. and Hamilton W.D. 1992). The next logical step would then be anisogamy, with one mating type developing into small, highly mobile sperm and the other into large, nutrient rich egg cells. Again, the process could have been driven by selection on fitness (Halliday T.R. 1994). In the end, it results in a situation that small, very mobile male gametes are in severe competition with each other to win the race for a big female gamete, rich in all that is necessary for the

132

successful development of an embryo. In other words, an interaction community with mutual selection has arisen.

In summary, sexual reproduction with genetic recombination and anisogamy could have emerged in unicellular eukaryotes for the following reasons. Genetic recombination corresponds to a key innovation that keeps the genetic profile broad by providing an additional barrier against a progressive specialisation under long periods of constant environmental conditions. As a result, the sexually reproducing unicellular forms could occupy niche situations in a sustainable way, which would not be possible for the generation sequences of solitary multiplying forms. For the same reason, the genetic recombination would also provide a benefit in the case of a catastrophic environmental event or change. Moreover, once isogamy has evolved, the development of anisogamy with mutual selection between two different types of gametes would be a logical next step resulting from selection on fitness.

From the *Interaction Theory* point of view, the emergence of sexually reproducing forms with meiosis and anisogamy would thus correspond to the next step to be expected, once unicellular eukaryotes have emerged. Further, the way in which this could have happened is in line with the law-like process of complexity increase (compare flowchart 1): a key innovation emerges in solitary multiplying form via a kind of symbiotic unification of cells. The new quality allows the occupation of a new type of niche and the emergence of a respective new phylogenetic cluster. It further provides a potential benefit with regard to survival from an environmental catastrophe. And last but not least, sexual reproduction goes along with a rising probability of multiplication that is visible in a development from quantity to quality. Thus, under favourable conditions, sexually reproducing microorganisms multiply slower than comparable asexual forms[8]. Their individual generation sequence requires therefore an on average higher likelihood to continue over time. Hence, the emergence of sexual reproduction corresponds to a complexity increase.

2.5. The evolution of sexually reproducing species

With the emergence of permanent interaction communities with mutual selection on interaction competence, the complexity increase in evolution starts into a new phase from the *Interaction Theory* point of view. From now on, the multiplication success of the individual form depends on its capacity to gain an interaction partner. As a consequence, the competition for multiplication success corresponds no

---

[8] For a general comparison of both kind of reproduction see e.g. Stearns S.C. and Hoekstra R.F. 2005.

longer to a general growth contest only. The new mutual selection rather allows a more balanced growth pattern, which makes possible a sustainable adaptation to niches at risk of becoming irreversibly exhausted or destroyed by exponential growth.

As a result, a world is left behind that is exclusively populated by solitary multiplying forms trapped in an ongoing rollercoaster between exponential growth and stagnation. From now on the journey of generation sequences towards increasing complexity can occur in irreversibly exhaustible or destroyable niches. This means that the multiplication of the respective forms depends on at least one niche factor sensitive to exponential growth. And as detailed in the first part, the sustainable occupation of these niches needs therefore permanent interaction communities with mutual selection on interaction competence, in order to overcome the general growth contest of solitarily multiplying forms. In the case of biological species, this mutual selection is mainly sexual selection. Without this self-made selection, complex ecosystems such as the tropical rainforest or coral reefs would not be possible. The transition from solitary multiplying form to sexually reproducing species means therefore that while the first depend on their capacity to use growth-resistant environmental situations as niches for their multiplication, the second depend on their capability to reduce the risk of unsustainable growth by mutual selection. The emergence of highly complex ecological networks would be a visible result of this capability. The reason is that it allows a more extensive coupling of the multiplication between different forms than is possible between solitary multiplying forms trapped in their general growth contest. While, however, the type of the niches is changing with the arrival of sexually reproducing species, the direction of complexity increase points still towards a rising probability of multiplication and continues to occur via the law-like process of complexity increase.

To support this view about species and sexual reproduction, I will in the following first present a new definition of species and then arguments as to why their evolution followed the law-like process of complexity increase and how this becomes visible.

The meaning of biological species

A commonly cited definition of the term species was first suggested by Ernst Mayr. By this definition species are "groups of actually or potentially interbreeding natural populations, which are reproductively isolated from other such groups". Today, different other species concepts are also in use, and no consensus on the exact definition and meaning of species has yet been reached (for critical discussion see Eldredge N. 1995 or Rose S. 1997). Here, *Interaction Theory* offers the following radically new view on the meaning of species:

- Biological species are reproductive interaction communities, which are able to occupy irreversibly exhaustible or destroyable niches in a sustainable way.

Hence, species should be understood as the entities of evolution occupying the second fundamental kind of niche that is not accessible for solitary multiplying form. With this, the above-quoted definition of species can be modified as follows. Species are groups of actually or potentially interbreeding populations, in which the multiplication of the individual depends on a specific kind of interaction competence, which isolates them reproductively from other such interaction communities. Thus a species consists of the sum of all individuals whose reproduction is influenced by the same kind of mutual selection on interaction competence. For this reason species would not be static. Local, geographically isolated groups can start changing their kind of interaction competence, which can result in the emergence of new species. This then, when the local interaction community has eventually modified its interaction competence in such a way that its members are no longer interbreeding with the original group.

Solitary multiplying organisms, such as prokaryotes, are often classified by their adaptations to their particular niche conditions (see e.g. White D. 2007). As a consequence, the characteristics and appearance of these organisms is strongly influenced by the qualities necessary for fast and efficient growth under their respective niche conditions. With the emergence of species this should radically change in the sense that the characteristics and appearance become now also strongly influenced by the respective mutual selection – the peacock is a concrete example. In other words, while the phenotype of solitary multiplying forms would be the result of selection on fitness, the phenotype of species would be the result of both mutual selection and selection on fitness.

Soma and germ cells as a result of the law-like process of complexity increase

Sexually reproducing organisms consist normally of so-called soma and germ cells (Campbell N.A. et al., 2008). Here, only the germ cells own the capacity to form gametes and to carry on the generation sequences. Thus in a figurative sense, the soma cells sacrifice their own potential generation sequences for the sake of the whole. The development of the first organism with soma and germ cell lines from a multi-cellular, undifferentiated precursor, corresponds therefore to a key step in the evolution of sexually reproducing species. A generally accepted explanation of why and how this has happened in evolution is, however, missing (see e.g. Maynard Smith J. and Szathmáry E., 1995). Regardless of how it has exactly happened, from the *Interaction Theory* point of view it represents a major complexity increase, which must have followed the law-like process of complexity increase.

In line with this view is the fact that the emergence of cell-differentiated organisms represents evidently a development from quantity to quality. Thus, an organism with a differentiation in germ and soma cells cannot under favourable conditions cause as many new generation sequences as an undifferentiated form of similar size, in which all cells can produce gametes. And for sustainability reasons, the correspondingly lower multiplication potential needs to go along with a rising probability of multiplication under comparable niche conditions.

Now, is the emergence of organisms with soma and germ cell lines also the result of the law-like process of complexity increase? The answer is yes, for the following reasons. In the case of a multi-cellular organism without a corresponding cell differentiation, all cells own in principle the potential to produce gametes. To conserve this potential, the cells would obviously experience restrictions regarding their differentiation for other purposes. In comparison, soma cells can show far-reaching differentiations and therefore provide the respective organism with new qualities, which are not possible for undifferentiated forms.

As an example, the soma cell differentiation allows the development of muscle cells, resulting in a much better mobility. As a result, the corresponding organisms could adapt to a new kind of niche that is not accessible for an organism of similar size and without a differentiation in soma and germ cells. In addition, it can be supposed that a lower mobility of multi-cellular organisms without cell differentiation may also represent a weak point for the probability of multiplication. This in the sense that they cannot effectively move to places with favourable conditions for releasing their gametes, so that the resulting zygotes have a higher chance to continue the respective generations sequence successfully. In other words, a superior mobility thanks to a soma cell line can increase the likelihood for the individual generation sequence to continue over time. And this in turn means that the differentiation in soma and germ cells can strengthen a weak point of the probability of multiplication.

In summary, thanks to a cell differentiation in soma and germ cells, sexually reproducing organisms would be able to adapt to a new kind of niche not accessible to undifferentiated forms. In addition, the cell differentiation can further increase the probability of multiplication, by improving the likelihood for the individual

136

generation sequence to continue over time. Both together are in line with the law-like process of complexity increase (compare flowchart 1). Hence, the emergence of organisms with soma and germ cells can be seen as the outcome of the law-like process, causing a rising probability of multiplication and thereby a further complexity increase during evolution.

The step from a situation in which all cells of a multi-cellular organism can produce gametes, to one in which only a restricted number of germ cells own this potential, is difficult to explain by an exclusive selection on fitness (see e.g. Wesson R. 1991). As mentioned several times, a pure selection on fitness forces the organisms to maximize their multiplication potential. This would obviously stand in conflict with the need to stop the multiplication of the designated soma cells. As a consequence it is difficult to imagine how a soma-germ cell differentiation can be driven by selection on fitness only. In contrast, from the *Interaction Theory* point of view the differentiation would correspond to a key innovation that requires dual selection. Hence, it would have been a coincidentally mutual selection that was responsible for why the cell differentiation started in the first place. In a following step, selection on fitness joined in and the cell differentiation got eventually fixed by adapting the organisms to a new kind of niche.

A consequence of this scenario is that the emergence of truly cell-differentiated organisms would not be possible without a reproductive interaction and the associated mutual selection. Now if this is true it means that the transition of undifferentiated to differentiated multi-cellular organisms can only occur within sexually reproducing forms. And indeed, this assumption is in line with what is known (see e.g. Wesson R. 1991).

Mutual selection and environmental adaptation of sexually reproducing species

As detailed in the first part, the sustainable multiplication of forms in niches at risk of becoming irreversibly exhausted or destroyed by exponential growth depends on mutual selection. In this regard, the negative coupling of the multiplication with other form(s) was identified as one reason why certain niche conditions are potentially sensitive to growth. This concerns in particular situations where a strong growth of one form can endanger the survival of another, as in the case of prey and hunter or host and parasite. Beside heterotrophic forms, this also concerns autotrophic forms, for instance as in the case of plants and herbivores or plants competing with each other for space, sunlight, etc. Now if this is true, it must be documented in the evolution of sexually reproducing species. In their evolution, the coupling of the multiplication with other forms must be of much greater importance than was the case in the previous phases of evolution. And indeed, this difference is well visible in the emergence of increasingly complex ecosystems, which followed the arrival of sexually reproducing species on the scene of evolution (see e.g. Heylighen F. 1996 or Knoll A.H. 2003).

It is generally agreed that sexually reproducing species are subject to two distinct kind of selection: the selection on fitness and sexual selection (see e.g. Stearns S.C. and Hoekstra R.F. 2005). From the *Interaction Theory* point of view, this dual selection allows species to overcome the general growth contest with its rollercoaster-like growth pattern. A good example that demonstrates how dual selection influences the population growth are species in which males fight for territories of a certain size and females only mate with males having conquered one of these territories (for corresponding examples see e.g. Halliday T.R., 1994). In this regard, it is necessary to mention the so-called group selection theory propagated by Wynne-Edwards (Wynne-Edwards V.C., 1962 and 1965). The idea that sexual and social selection have the role of preventing the population from exhausting the recourses is at the centre of Wynne-Edwards' thinking. Obviously, this understanding is in accordance with above argumentation. On the other hand, group selection theory was built on the assumption that groups are a distinct unit of natural selection. This view was heavily criticized, by insisting that the individual corresponds to the main target for selection (see Williams G.C., 1966). From the *Interaction Theory* point of view, both positions can be reconciled. Here, the individual is the target of both selection on fitness and mutual selection. However, only the generation sequences of those interaction communities which own a mutual selection that causes a sustainable multiplication in the respective niche can continue over time.

Now, if the arguments regarding the role of mutual selection are correct, it is to be expected that under certain conditions one type of selection is more influential for the evolution of a species than the other. Let's take for instance places with frequent climatic changes, where the fate of the generation sequences depends strongly on the capability of the individual form to survive the environmental challenges. Thus, selection on fitness should be dominant under these conditions. In contrast, the situation must be different in the case of very stable and reliable conditions that are favourable for growth. Here, the need for a strong mutual selection must be evident, because the fate of the generation sequences would be dependent on the capability to act against unsustainable growth.

These conclusions correspond to what is observable in nature. Ecosystems such as the tropical rainforest or coral reefs are good examples. The overall physical conditions in both habitats are in general exceptionally stable and may in principle allow undisturbed growth over many consecutive generations [9]. Yet the population density of most species in both ecosystems is rather sparse and a sudden explosion in the number of one or a few forms does normally not occur (at least not to my knowledge). From the *Interaction Theory* point of view this would only be possible because of a strong mutual selection within the respective species, which acts as a barrier against a development towards a general growth contest. And indeed, this is in line with the fact that both ecosystems are characterized by an impressive luxury with regard to colours and forms. This splendour would document the consequences of the strong mutual selection, which corresponds mainly to sexual selection as well as the selection between flowering plants and pollinating insects. In contrast, the ecosystems in the northern hemisphere, where the overall climatic conditions are much more variable, do not show the same richness in colours, form and biodiversity. In addition, certain forms can appear here in very high numbers (Eldredge N. 1995). Again this is in line with above assumption that under challenging conditions, the fate of the generation sequences is much more influenced by selection on fitness. As a consequence, the species in these habitats look also 'more boring'. As a result, the observable differences between different habitats support the conclusion that it depends on the environmental situation whether mutual selection on interaction competence or selection on fitness is more relevant for a particular species.

In the first part I have connected mutual selection on interaction competence with the emergence of new species. The reason is that an isolated interaction community can develop a new kind of mutual selection, which may over time result in their reproductive isolation and thus a new species. This means that in habitats where mutual selection needs to be strong, new species should appear more easily and the biodiversity would therefore be higher. Consequently, the biodiversity in habitats with in comparison very stable environmental conditions, such as the tropical rainforest or coral reefs, must be particularly high. And obviously, this is precisely the case (see again Eldredge N. 1995). Here, new species develop apparently more easily than in places where the individual multiplication is much more influenced by selection on fitness than by mutual selection.

The importance of mutual selection for the sustainability of generation sequences under environmental conditions which are in particular sensitive to growth is also supported by the following observation. Vbra has remarked that specialists differ more than generalists from the original morphological norm of the respective genus (Vrba E.S., 1984). This means that ecologically specialized species possess

---

[9] This is the reason why these habitats are in particular in danger of becoming destroyed by human activity (for the diversity in the tropical rainforest and its current destruction see e.g. Wilson E. O. 2010).

many anatomical features or adaptations that are themselves evolutionary special-izations. And most of these specializations cannot simply be explained as adaptation to their specific resource situation. Compared herewith, ecological general-ized species tend to retain many of the primitive features that were present in re-mote ancestors. Again this can be explained by the relative importance of mutual selection compared to selection on fitness. Specialists are normally entirely de-pendent on one resource, which makes them relatively vulnerable. This is con-firmed by palaeontology showing that specialists become more frequently and rapidly extinct than generalists (see e.g. Eldredge N. 1995). As a consequence of their more vulnerable resource situation, it can be expected that specialists have, compared to generalists, a greater need for a strong mutual selection on interaction competence. It would provide the necessary barrier against a development to-wards unsustainable population growth by a progressively mounting fitness. And obviously, the stronger influence of mutual selection can explain the relatively more anatomical changes compared to generalists and the fact that these features are not simply an adaptation to the specific resource situation, as was observed by Vbra.

Complexity increase and the intensification of mutual selection within species

In the following I will present a further example of how the basic principle of *Interaction Theory* with its law-like process of complexity increase becomes vis-ible in the evolution of sexually reproducing species. In this case it concerns the insight that the biological complexity increase in niches at risk of becoming irre-versibly exhausted or destroyed by exponential growth requires an intensification of mutual selection, in order to be sustainable.

Well, why must the mutual selection in the course of evolution become more in-tense? It was noted that selection on fitness is forcing multiplying forms to max-imize their number of offspring under the respective environmental conditions. As a consequence, selection on fitness alone provokes a general growth contest. In irreversibly exhaustible or destroyable niches, the forms need therefore repro-ductive interactions. The resulting mutual selection can ensure a sustainable bal-ance between environment and multiplication, by acting against a general growth contest. Moreover, it was noted that with each round of the law-like process of complexity increase, the likelihood of continuing over time rises for the corre-sponding individual generation sequence (see flowchart 1). Thus, at the end of each round fewer sequences are stopped under comparable environmental condi-tions, compared to the previous situation. To ensure the sustainability of the cor-responding generation sequences in the irreversibly exhaustible or destroyable

140

niches, their branching frequency must therefore decrease, according to a development from quantity to quality. This in turn means that the same kind of mutual selection, which was adequate at the start of a new round in the law-like process, would no longer be sufficient at the end. Hence, mutual selection must become in some way more intense in the course of the successive rounds of the law-like process - this refers to forms of increasing complexity, which multiply under similar conditions. This intensification must be visible in the evolution of sexually reproducing species.

The mutual selection on interaction competence within species can in principle occur via different criteria such as olfactory, visual, acoustic or tactile. The intensification can therefore mean that more complex species use more of these criteria. Furthermore, it can mean new extra forms of interaction which emerged in addition to sexual reproduction, such as social interaction. These kinds of differences should exist between related groups of the same phylogenetic line, as for instance between reptiles and birds. The taxonomic group emerging later in evolution, here the birds, should own a greater intensity and variety of mutual selection than the older group in the corresponding phylogenetic line, here the reptiles.

The assumption corresponds to what can be observed in nature. So birds, in comparison with reptiles, obviously own the richer mutual selection, with particularities such as the birdsong and other kinds of sophisticated, bird-specific features related to mating. Very impressive examples in this regard are the astonishing bowler birds (for this and other respective examples from birds see Rothenberg D., 2012). For mammals the same can be said, compared to reptiles, in particular if social selection is included. Chimps and dolphins are here the evident examples, particularly because of their richness in social interaction (see e.g. Byrne R.W. 1994). To my knowledge, a comparable degree of mutual influence on the single individual and its multiplication can be found nowhere else in the phylogenetic tree – obviously with the exception of *Homo sapiens*, representing the tip of the iceberg.

An intensification of mutual selection by a new and additional kind of interaction has also occurred in the evolution of terrestrial plants. It concerns gymnosperms in comparison with the phylogenetic more recent group of angiosperms. As a result of their interaction with insects, the multiplication success of sexually reproducing angiosperms is also dependent on their ability to attract the respective insects for collection and distribution of their pollen (see e.g. Soltis D.E. et al. 2008). This is obviously not the case for gymnosperms. Hence it can be said that angiosperms are the subject of a more intense mutual selection.

The same argument is valid with regard to eusocial insects, such as ants and bees. Here, the multiplication depends on the new interaction between the different castes, which is not the case for other insects. In this regard it is interesting that the reproductive interaction of eusocial insects is characterized by the particularity that in the female line only the queen is part of the continuing generation sequences (for the genetics of eusocial insects compare e.g. Maynard Smith J. and Szathmáry E., 1995). In this respect, the interaction is comparable with the one between soma and germ cells or between protein and DNA.

In conclusion it can be said that the mutual selection becomes indeed more intense with increasing complexity, as is to be expected from the law-like process of complexity increase. From the IAT point of view, the more intense mutual selection ensures the sustainability of the multiplication. And this is necessary, since with increasing complexity the likelihood of the individual generation sequence continuing over time steadily rises.

Species with asexual reproduction

Most but not all higher organisms reproduce sexually, and especially within plants asexual reproduction is not uncommon (Wesson R. 1991). Why is it that not all species need sexual reproduction, if it is the means for a sustainable occupation of niches irreversibly exhaustible by growth? A possible explanation of why the generation sequences of certain species can continue over time without this reproductive interaction might have to do with the respective niche situation.

For unicellular microorganisms, the fact that certain forms reproduce sexually and others not was connected with genetic recombination (see previously). The latter can prevent an ongoing specialization of the genetic profile. This would be beneficial in an environment that is very stable over manifold generations, as in the previously given example of sexually reproducing algae. Here it was also concluded that in niches with a sufficiently varied selection by the environment, the genetic profile of asexually reproducing microorganisms is not at risk of an overspecialization. As a consequence, the generation sequences of these forms are sustainable without genetic recombination.

In addition to genetic recombination, a key insight of *Interaction Theory* is that when it comes to sexually reproducing higher organisms, it is also the mutual, respectively sexual selection which is required for a sustainable niche adaptation. In analogy with micro-organisms, it can therefore be assumed that asexually reproducing higher organisms possess a particular niche situation. This means that either the respective environmental selection is sufficient, so that the risk of unsustainable growth is so low that there is no need for mutual selection; or, alternatively, the respective niche tolerates strong growth without the risk of becoming irreversibly exhausted or destroyed. Irrespective of the precise reason, it is to be expected that asexual reproduction should in general be more frequent in situations where mutual selection is normally less intense. As outlined above, the mutual selection intensifies with increasing complexity, and less complex species require in general less mutual selection. Thus, asexual reproduction should be more frequent in species with lower complexity, which is actually the case (for asexual reproduction see e.g. de Meeûs T. et al. 2007).

Another conclusion that can be drawn is the following. If complexity increase in evolution and the development of the clustered biodiversity is the result of the law-like process of complexity increase, it means that asexually reproducing species must descend from sexually reproducing ancestors. The reason is that the forms could not acquire a key innovation without mutual, respectively sexual selection. Consequently, asexual species would not cause a new phylogenetic cluster and only evolve within existing clusters. Actually, this conclusion is in line with the generally accepted view that asexual species have developed from progenitors with sexual reproduction (see e.g. Wesson R. 1991).

The link between the reproduction type and the particular niche situation also offers an explanation of why asexual reproduction is more common among plant species than among animals (for on overview about asexual species see de Meeûs T. et al. 2007). As autotrophic forms plants utilize as key resources CO_2 and sunlight, which do not become irreversibly exhausted by growth. Therefore it can be expected that their niches are on average less sensitive to an irreversible exhaustion by growth than those of the heterotrophic animals. Hence, the higher frequency of asexual reproduction among plants.

2.6. Species with social interaction

The central insight of *Interaction Theory* is that biological complexity goes along with a continuously rising probability of multiplication, documented by a development from quantity to quality. Hence, biological complexity increase signifies that the individual generation sequence has an on average rising likelihood to continue over time, while showing in parallel less branching events per time. As outlined in a previous chapter, for species adapted to niches irreversibly exhaustible or destroyable by exponential growth, this demands that their ongoing complexity increase is accompanied by an intensification of mutual selection. In line with this conclusion is the fact sexually reproducing species developed indeed additional kinds of interactions impacting their reproduction, such as between angiosperms and insects or in the case of mammals between the members of social groups. Here, the respective individuals are exposed to an additional, new type of mutual selection. In the following I will discuss social interaction under this aspect, and why it made a further rising probability of multiplication possible. As the reader will see, this will lead directly towards the evolution of *Homo sapiens*, which I will discuss afterwards in the last chapter.

Social selection and its consequences

In previous chapters I have differentiated between sexual and social selection as two distinctive forms of mutual selection within species. Now, what exactly is the difference between sexual and social selection? The term sexual selection, as used in this book, stands for phenomena that are directly related to mating, such as mating behaviour and sexual ornamentation. In other words, the sexual interaction is mainly the competition within the same sex with regard to mating and reproduction. In contrast, social relates to the interactions between all members of a social group, which do not necessarily have a direct connection with mating and reproduction as such (for social groups see e.g. Lee P.C. 1994). Hence, social interaction includes also the cooperation between members of the same sex as well as the competition between the two sexes.

It is essential for social interaction that the respective group shows some kind of social hierarchy and that the position or rank within this hierarchy influences the individual multiplication success (see e.g. Lee P.C. 1994). As a consequence, social interaction must always be seen in the context of a particular social group. Decisive from the *Interaction Theory* point of view is that social interaction causes a new form of mutual selection, which can be called social selection. To be successful in this regard needs obviously specific, individual qualities that may be very different from those relevant for sexual selection. And because these socially relevant qualities or competences vary between the individuals in a group, some are socially more successful than others. In the end it means that social selection results from the fact that the multiplication success becomes influenced by the acquired social status of an individual within its group. The differences in the individual social competence become therefore important for the mutual selection within respective groups.

The simple fact that social interaction takes place between the individuals of a group does not automatically mean that all animals, living permanently or seasonally together, interact socially in the above-mentioned sense. For example, large herds can be simply hold together by the benefit for the individual of being better protected against beasts of prey (Lee P.C.. 1994). For this reason, the members might show not much more interaction than those associated with sexual reproduction. The seasonal fighting between males in large herds that is related to mating would be an example. Compared herewith, in groups with the described social interaction, the individuals interact with each other permanently. Thus, the members of a social group show forms of cooperation, such as sharing the responsibility for the young or joint hunting.

In the real world the clear distinction between sexual and social selection may not always be easy. So, both types of selection are strongly interconnected, as can be clearly seen in social groups of chimps or dolphins. This should not come as a surprise, since social selection came later and had to build on the already existing sexual interaction. Yet, the differentiation between both forms of interaction is important with regard to complexity increase. The reason is the following. Sexual selection is normally related to inherited qualities which influence the respective interaction competence of the individual. Examples are the possession of gender-related attributes in order to be accepted as a mating partner, such as a penis of a specific form and size in many invertebrates, or an impressive antler in the case of a stag, or the splendid colour and appearance of certain birds (for a discussion of gender related attributes and selection see Wesson R. 1991). The differences in these qualities, which determine the individual interaction competence, and thus the multiplication success, are normally caused by genetic variations. In other words, sexual selection is mainly driven by differences in the genetic profile between the individuals of a respective interaction community. In contrast, social interaction competence needs to a large extent to be acquired by learning, even if genetics may also play a role (for learning and evolution see e.g. Byrne R.W. 1994). And it is because of this difference with regard to learning that it is important to distinguish between both forms of interactions. As I will show below, this becomes decisive for the discussion about the rise of *Homo sapiens*.

The relation between learning and social competence can briefly be summarized as follows: the consequence of social selection is that individuals with a high social competence can conquer a high rank in the social hierarchy. And the high social status can then translate into a relatively greater multiplication success, compared to the lower ranking members. The corresponding differences in the social competence are caused by physical as well as behavioural attributes. For example, the ability to form alliances with others is a behavioural quality with a strong influence on the social status of a chimpanzee. The corresponding behaviour is to a large extent learned or taught during the youth (e.g. Byrne R.W. 1994 or Bonner J.T. 2017). Thus, the social competence of an individual depends on postnatal learning, which is not, or to a much lesser extent, the case with purely sexual attributes.

Now, if the interaction competence is suddenly not only influenced by hereditary disposition but also by postnatal learning, this must have visual consequences. An investment into the learning of social competence during youth would become beneficial, because it can pay off in a relatively higher multiplication success. Obviously, this gives the youth phase a completely new importance and can explain the striking prolongation of this phase within certain mammal species. The longer youth allows the learning of the necessary competences for later life.

Supporters of conventional evolutionary theory will probably agree that the prolongation of the youth phase in certain animal species allows the learning of social competence. Nevertheless, they may insist that the development towards a longer youth period is the result of selection on fitness, because it makes the respective species also fitter. But how realistic is this? Clearly, the learning period during youth may in the end cause benefits for the environmental fitness. Considering, however, the multiple difficulties and problems related to the delicate youth phase it is hard to imagine how its progressive prolongation could solely be driven by fitness benefits. This in particular, because the longer youth means a prolonged time until sexual maturity, which will limit the overall number of possible progenies during a lifetime. The solution offered by *Interaction Theory* is once more that of a dual selection – see below.

Rising probability of multiplication by social interaction

The key message of this book is that complexity increase during evolution occurs by a law-like process towards a progressively rising probability of multiplication, visible in a development from quantity to quality. Now, if social interaction results in a further complexity increase of species, it means that socially interacting forms must have an on average higher probability of multiplication, compared to their phylogenetic ancestors. If these species are characterized by a prolonged youth phase this is obviously the case, simply because it means a long investment into the single offspring that reduces inevitably the multiplication potential. For example, highly social species such as chimps, dolphins or elephants are characterized by such a relatively long period until they become sexually mature. Thus, they can only have less offspring per time, compared to an animal of similar size, but without a corresponding investment into the youth phase. As compensation, the individual generation sequence of these highly social species requires a relatively high likelihood to continue over time; otherwise they risk becoming extinct.

Following the logic of *Interaction Theory*, the complexity increase by social interaction must also occur via the law-like process of complexity increase. The key innovation would be in this case the brainpower that is necessary to learn social and environmental skills – in the next chapter I will come back to this point in more detail. As for all key innovations, the emergence of brainpower would also be caused by a dual selection that starts with mutual selection in a corresponding interaction community. Hence, the initial reason why brainpower starts to develop during youth would be its positive impact on social competence, and thus on the multiplication success. Over time, the corresponding investment would also result in the capacity to cope with environmental challenges in a new way. Latest at this point, selection on fitness joins in. The resulting dual selection would then drive the development toward species which learn, during a prolonged youth phase, how to master social and environmental challenges. This again would allow the occupation of a new kind of niche, not accessible without the respective brainpower. And finally, at the moment the established ecosystems become destroyed, the capacity to deal in this new way with challenges would drive the further complexity increase – see next chapter.

A consequence of the dual selection described would be that the acquisition of both social competence and environmental fitness should be closely linked and mixed. That this is indeed the case is well visible in social mammals, such as elephants. If these animals have grown up in captivity, they are often not able to survive in the wilderness. This corresponds to a visible sign that they did not learn the necessary skills how to survive and raise their progeny in the natural habitat. In addition, highly social animals such as elephants, having grown up in captivity, are normally also not well accepted by their wild conspecifics (for the reintroduction of animals grown up in captivity, see e.g. Laidlaw R. 2001). Personally, I am not aware of other phylogenetic groups than mammals and maybe certain birds, for which this is the case. Anyhow, it shows nicely that in socially interacting animals the learning of social competence and environmental fitness is closely interconnected, as can be expected of a dually selected key innovation.

To further illustrate the interconnection between social interaction and environmental fitness, I will use once more the example of elephants. Their trunk corresponds without doubt to an essential feature of this species. It is central for many different aspects of their life, be it for survival in the environment or for sexual and social interactions. As already mentioned in other cases above, for corresponding, sophisticated organs or structures it is difficult to imagine how their emergence during evolution can solely be the product of selection of fitness. For the developing trunk it would mean that it has to deliver instantly a sufficiently strong and over multiple generations continuous benefit for the individual fitness, so that it can develop in its final form. In this particular case it is even more challenging, because the benefits do not only depend on the trunk as such, but also on the need to learn its respective use during youth. For these reasons it again makes sense to assume that the original emergence of this structure occurred via mutual selection. Given the central importance of the trunk for elephants, the corresponding mutual selection might have been of sexual as well as social nature. In a following step, selection on fitness would then have jumped up and gradually improved the utility of the developing trunk also for environmental purposes. With this dual selection the elephant trunk could eventually develop into its final form and function. As a result, the organ must play a role for the sexual and social competence as well as for the daily struggle for food and water, as is indeed the case (for elephants see e.g. Kangwana K. 1996).

Not all animals with a prolonged youth and learning phase live in social groups. However, it was said above that the initial reason why the youth phase has become significantly longer would be the consequence of mutual selection on social competence. The acquisition of new skills relevant for the environmental fitness would then have followed. Consequently it would mean that the non-social species with prolonged youth phase can only have developed from socially interacting ancestors. During their evolution they then adapted to a less social lifestyle, but the species would still depend on the learning during its youth. That this is indeed the case remains to be confirmed by respective experts.

In summary, from the *Interaction Theory* point of view, highly social mammal species with a prolonged youth phase represent in their branch of the phylogenetic tree a further step towards a rising probability of multiplication through the continued development from quantity to quality. In the next and last part, I will discuss the endpoint of the corresponding development, namely *Homo sapiens*.

2.7. The evolution of *Homo sapiens*

In this section I will discuss the most recent phase in the journey of generation sequences travelling through deep time towards a rising probability of multiplication. This phase corresponds to the history of ourselves, *Homo sapiens*. The complexity increase that took place during this phase is largely associated with an increasing brain complexity and a resulting increasing brain power (for the human evolution see e.g. Tattersall I. 1998). The phase would start where the previous ended, with species characterized by multifaceted social interactions and a prolonged youth phase that allows the development of brainpower and herewith the learning of socially and environmentally related skills. In this respect, *Homo sapiens* corresponds obviously to the tip of the iceberg. No other living being has a comparable degree of social interaction and an equally long and vulnerable youth phase, in which the individual acquires the abilities for further life. From the *Interaction Theory* point of view however, our evolutionary history has an additional particularity. It is closely connected with a fundamental change in the over-all niche situation, as I will outline below. In brief it means that the outstanding evolutionary success of our species can be linked to the capability to overcome the niche boundaries valid for other biological forms. Here lies the reason why the rise of *Homo sapiens* would mark the start of a new phase of complexity increase.

When it comes to our own species, probably the most interesting question from the evolutionary point of view is: does evolution inevitably result in an intelligent being such as *Home sapiens*? The answer given by *Interaction Theory* is yes: biological complexity increase during evolution goes into the direction of an intelligent being. The reason is that from a certain point on, the probability of multiplication can only rise further by intelligent behaviour. Only with brainpower it is possible to adapt to the environment in a new way, so that a further rising probability of multiplication, and thus complexity increase, becomes possible. Consequently if enough time is given, the ongoing development towards a rising probability of multiplication will cause at some point the emergence of an intelligent form which can adapt to the environment with its brainpower and thus start a new phase of complexity increase. In the following I will deliver arguments to support this insight.

Before however, I want to mention once more Simon Conway Morris, who came with a different approach to the same conclusion: social species with highly developed brains must be seen as inevitable in evolution. His thinking is based on the argument about the central role of convergence, instead of contingency, and the observation that evolution in general has a trend towards what he calls "mammalness" (Conway Morris S. 2003). Obviously, this conclusion would be in line with the central theme of this book that evolution is going towards a rising probability of multiplication.

The brain and the probability of multiplication

How can the probability of multiplication continue to rise after evolution has reached the stage of species with social interaction and a prolonged youth phase? To answer this question, I want first to summarize the conclusions of the last chapter about complexity increase and social interaction: as a consequence of social selection the multiplication success of an individual is influenced by its social competence, which must be learned and trained in a prolonged youth phase. The corresponding quality which makes this possible was identified as brainpower. At the moment when the consequences of the social interaction also impact, directly or indirectly, survival, selection on fitness joins in. As a result of the dual selection, brainpower becomes a new key innovation that allows the occupation of a new kind of niche within the established ecosystem. A following large-scale catastrophic event or change drives then a new round of the law-like process of complexity increase, because the brainpower helps the corresponding species to survive relatively better and thus to become dominant in the new ecosystems (see flowchart 1). Based on this scenario, I will discuss first in more detail the connection between social interaction, environmental fitness and brainpower.

A sophisticated social environment is highly dynamic. In social communities the group dynamic and relations between the individuals can suddenly change and confront the members with new, challenging situations. Chimpanzees and dolphins are examples of how demanding and stressful social interaction can become for an individual. In a respective environment, a member can only achieve a high social status by owning the capability and flexibility to deal with the multiple, often unpredictable challenges better than the others. Obviously this would result in the positive selection of individuals who are able to react adequately to these challenges or can even anticipate them. The tool to do so is clearly brainpower and a corresponding intelligent behaviour. Not surprisingly species with complex social relations, such as chimps, elephant or dolphins, have highly developed brains. As a consequence, the development of brainpower is seen by many experts as a result of complex social interactions. This again resulted in the formulation of the so-called Machiavellian Intelligence hypothesis, which says that social interaction is the catalyst for the development of intelligent behaviour during evolution (for more details see e.g. Byrne R.W. 1994 or Bonner J.T. 2017). And without doubt, the Machiavellian Intelligence is helpful not only for social status, but also for surviving environmental challenges. This understanding corresponds to what was said before about social interaction and the development of brainpower. With *Interaction Theory* it becomes, however, part of a wider evolutionary context, namely the law-like process of complexity.

For biological complexity increase, as described in this book, it is necessary that a new key innovation has a positive impact on the generation sequences of the corresponding species. It needs to result in a greater likelihood for the individual sequence to proceed over time, compared to those of similar forms without key innovation and comparable niche conditions. For large-brained, intelligently behaving species this means that the probability of multiplication of their generation sequences must rise on average in the ecosystems that newly develop after a large-scale environmental stress. For sustainability reasons, a respective rise must go along with less offspring per time, in the sense of a development from quantity to quality. As already said, this is indeed the case with large-brained, highly social species such as chimps or dolphins. They have relatively fewer offspring per time, compared to other species of similar size and without comparable social interactions[10]. The reason for the prolonged youth phase is obviously that a large brain as such is of no use and rather very costly in terms of energy (see e.g. Tattersall I. 1998). The large brain becomes however an advantage if the owner has sufficient time to train the herewith possible brainpower. Thus, an increasing brainpower goes hand in hand with a rising investment into the youth phase and consequently a longer generation time. The need to train the brainpower during the youth phase makes it therefore the perfect quality to further drive the development from quantity to quality.

Now let's turn to the question of how exactly the progressively increasing brainpower can cause a further rising probability of multiplication. In the first part I have identified the early phases in the lifecycle of higher organisms as the decisive weak point for their generation sequences. The risk that a generation sequence gets stopped is normally much higher in the delicate youth phase than during adulthood. This should in particular be the case for highly social forms with a prolonged youth phase. As a consequence, the answer to the above question would be that the increasing brainpower can drive the further rising probability of multiplication via a resulting better protection and care of the offspring. This conclusion, together with the principles of the law-like process of complexity increase (compare flowchart 1), makes it possible to deduce the following scenario: in a social interaction community, the dual selection continues to strengthen the brainpower, which leads eventually to better care and protection of the offspring. This in turn allows the adaptation to a completely new niche situation. With regard to the evolution of the *Homo* genus this could for instance mean that more brainpower provided efficient access to meat as a new nutritional resource (see again Tattersall I. 1998). It is apparent that the capacity to exploit such a rich nutritional source results is a significant benefit for the development during the youth phase. And it makes further sense that this also helps to survive a sudden large-scale catastrophe better than a comparable species without the capability to exploit this nutritional resource efficiently. Regarding the meat example, it should also be mentioned that in the evolution of the *Homo* genus a nutrition rich in protein was essential for the development of larger brains (see e.g. Leakey R. 1994).

After an intelligent, large-brained species survived a catastrophic event or change significantly better than comparable forms, it can play a more important role in the newly developing ecosystems (compare flowchart 1). As a result, the corresponding species can now also take possession of niche situations which were

---

[10] This is simply the consequence of the fact that an increasing generation time, due to a longer youth phase, decreases the number of possible progenies per generation, compared to the original situation.

previously occupied by others. However, as a consequence of the greater protection and care in their youth phase, the respective species would continue the development from quantity to quality, to ensure a sustainable balance with the environment. And this means that the adaptation to corresponding niche situations goes along with an additional investment into the single offspring, causing a further prolonged youth phase. This again would lay the basis for the next round of the law-like process. In the end, the brainpower increases stepwise with each round, resulting in an even longer youth phase that is characterized by a progressive learning of sophisticated, socially and environmentally relevant skills.

In summary, it is well supported that social interactions can drive an investment into brainpower. And a rising brainpower needs not only larger brains but also sufficient time and the right conditions so that the individual can learn to use it effectively. The emergence of species with highly developed brains and a long learning period before they reach sexual maturity demands therefore a development from quantity to quality for a rising probability of multiplication. Before I compare these conclusions with what is known about the evolution of *Homo sapiens*, I want first to discuss a decisive consequence of an increasingly intelligent behaviour.

Brainpower and niche adaptation

From the *Interaction Theory* point of view, brainpower would in principle just be another key innovation, like those previously discussed. On the other hand, a decisive difference exists. Compared to the preceding key innovations during biological evolution, brainpower is special with regard to environmental adaptation for the following reason. The resulting intelligent behaviour makes it possible to deal with new situations and challenges in an entirely new way.

In the course of biological evolution, the specialisation of species on particular environmental situations is normally connected with the acquisition of anatomical or morphological features, such as a strong beak to crack nuts, or a long and fine muzzle to feed on hidden insects. In other words, the fitness as well as the interaction competence of the individual form depends largely on its morphological qualities and structures. Evolution and complexity increase resulted therefore from the fact that selection on fitness as well as mutual selection can choose between the available variations of these qualities and structures. With the emergence of brainpower and an increasing importance of intelligent behaviour for the individual interaction competence and fitness, this would change. Now it becomes possible to adapt to environmental conditions in a different way. An example would be the exploitation of new food resources solely by the intelligent use of tools and techniques, such as the use of stones for cracking nuts or little branches for collecting insects, e.g. by chimps (see e.g. Byrne R.W. 1994). Thus with growing brainpower, the environmental adaptation becomes increasingly a consequence of intelligent behaviour and depends no longer solely on the acquisition of specific anatomical or morphological features. For the law-like process of complexity increase this must have consequences. Now it no longer depends only on the lengthy emergence of new morphological and anatomical key innovations over multiple generations. It can now also be driven by key innovations that result from intelligent behaviour, which should make the process much faster - I will get back to this point.

The possibility of adapting with brainpower in a fast and flexible way to new environmental situations must also have consequences for the niche adaptation in general. Thus, for a corresponding species with exceptional brainpower the niche boundaries would lose their previous importance. With niche boundaries I mean the fact that species cannot normally switch easily from one kind of niche to another, in particular if this requires specific morphological adaptations, such as a strong beak. Respectively, niche-specific qualities need to be acquired in a rather slow process over multiple generations. This can be seen as a kind of boundary which protects the species adapted to the particular niche to a certain extent against an easy ousting by others. In contrast, an exceptional brainpower would in principle allow a much faster adaptation to new environmental situations, e.g. from one generation to the next. As a consequence, for a highly intelligent species the niche boundaries would no longer exist in the same way. This allows the conclusion that if such a species can overcome the so far existing niche boundaries, it must have far-reaching consequences. Evolution and herewith complexity increase would enter into a completely new phase that I will call the unrestricted niche phase of evolution. This term expresses the fact that from a certain point, the growing brainpower will result in the capability of adapting to new environmental situations in an unprecedented fast and flexible manner (for the uniqueness of human evolution see Tattersall I. 1998).

In summary, brainpower makes the adaptation to new environmental situations in an unprecedented way faster and more flexible, because it no longer only depends on the relatively slow acquisition of new morphological or anatomical qualities. This would allow a highly intelligent species to overcome the traditional niche boundaries and to start a new phase in evolution that can be called the unrestricted niche phase. When this happens it should obviously have visible consequences, which I will discuss in the next section.

Entering the unrestricted niche phase of evolution

Based on the above considerations, the course of generation sequences towards a rising probability of multiplication will, if enough time is given, result in an increasing brainpower that leads finally into what I call the unrestricted niche phase of evolution. Next, I will make predictions as to how the entry of an intelligent species into the unrestricted niche phase of evolution would occur from the *Interaction Theory* point of view. Afterwards I will then compare the assumptions made with what is known about the evolution of *Homo sapiens*.

The entry in the unrestricted niche phase would follow the law-like process, such as the previous steps of complexity increase during evolution: a new, highly social species with unique brainpower emerges within an established ecosystem as the result of dual selection. The particularity of this species is that, thanks to its intelligence, it has the capacity to learn in an unprecedented manner new skills relevant for survival and fast adaptation to the environment. Therefore it can adapt to environmental situations in a way not possible for other species. Over time a large-scale environmental catastrophe occurs and results in the destruction or massive disturbance of the existing ecosystems. Thanks to its unique brainpower, the corresponding species survives better than other comparable but less intelligent forms and as a result starts to play a much more dominant role in the aftermath. This means that it can now adapt to a multitude of environmental situations which were previously occupied by other forms. In evolutionary terms this occurs extremely fast compared with previous events of this kind. This rapidity is due to the fact that the adaptation to new situations can now occur mainly through intelligent behaviour and much less through morphological adaptations. In the events following large-scale environmental stress, the adaptation of the key innovation-owning species to new environmental situations should therefore no longer result in the development of a new biodiversity, as was the case up to now. The form that enters the unrestricted niche phase should rather show a kind of diversification that is mainly based on behavioural differences. In other words, the entry into the new phase should be marked by the emergence of a cultural diversity.

The key message of this book is that complexity increase results in new species with a relatively higher probability of multiplication. Obviously the species entering the unrestricted niche phase should mark a new highlight in this respect. It should own the so far highest probability of multiplication in evolution, because its individual generation sequence would have an on average unmatched likelihood to continue over time. For sustainability reasons, the species would at the same time also be characterized by a further intensification of mutual selection that exposes the individual in its social group to an unprecedented level of this self-made selection, compared to other living forms.

154

Within established ecosystems, the species are adapted to their specific niches. The sudden invasion of niches by other species is therefore more an exception than a rule (see e.g. Pianka E. R. 2000). As already indicated, at the moment a highly intelligent species enters the unrestricted niche phase, this should change. It is to be expected that the highly intelligent form drives out other species from their established niche, by overcoming the traditional niche boundaries and exploiting all kind of environmental conditions. This implies also that conflicts should become frequent between the different groups forming the diversity of this species. The entry of a highly intelligent species into the unrestricted niche should therefore not only become visible in the sudden extinction of other species, but also in growing conflicts between its socially and behaviourally different interaction communities.

The rise of *Homo sapiens*

Are the above scenarios in accordance with reality? For this, let's first ask whether what is known about the evolution of *Homo sapiens* corresponds indeed to the principles of *Interaction Theory*? The answer is yes for the following reasons.

Evidently, the rise of *Homo sapiens* is a result of its unique brainpower and the associated capability to acquire skills that are essential for survival. In this regard it is striking that the fossil record which documents the evolution towards anatomically modern humans does not show a gradual growth in brain size. It looks rather like a stepwise development (e.g. Leakey R. 1994 or Tattersall I. 1998). This would be consistent with the law-like process of complexity increase, in which each further boost in brainpower has to start in a reproductively isolated interaction community. As discussed, these local groups would be rather small and therefore normally not visible in the paleontological record. At the moment however, when the increasing brain size resulting from the corresponding social selection provides eventually also a decisive fitness benefit, the local group can start to spread out by adapting to a new kind of niche. As a consequence, the chance of finding them in the form of fossils starts to rise. The fitness benefits of an increasing brainpower in the Homo lineage that could have driven this process are the mastery of fire and the making of effective tools for hunting and meat processing. After each round in the law-like process, the further development had to wait until once again an isolated interaction community with strong social selection started a new round. In this way, the development towards larger brains should appear as a stepwise development in the fossil record, as is the case. For example, *Home erectus* with an already relatively large brain – only about one third smaller than ours – seems to appear suddenly on the stage of evolution about 1.7 million years ago. Then the brain size did not change significantly for a long time. In a similar way, about 450,000 years ago the first Neanderthals become detectable, with a brain size slightly above ours. And finally, about 200,000 years ago, *Homo sapiens* appear for the first time (for human evolution see again Leakey R. 1994 or Tattersall I. 1998).

The fact that Neanderthals had about the same or even slightly larger brains than us illustrates nicely that it is not only size that matters. It is rather necessary to learn how to use this sophisticated organ successfully. Again this is in line with the assumption made about brainpower and the role of the social environment.

The evolution of brainpower is once more a good example that it is easier to explain the sudden emergence of a key innovation by dual selection than solely with selection on fitness. In this particular case the explanation given by *Interaction Theory* helps to avoid the following paradox: if brainpower evolved only because the progressive increase in brain size provides continuously fitness benefits, we would expect that it is strongly favoured by natural selection on fitness. Now, if this were indeed the case, why then do more species not become highly intelligent? This is in particular surprising because it is relatively easy to identify strong fitness benefits for most animals, if only the beasts would behave more intelligently. Nevertheless, only relatively few species show a significant investment into brainpower (Byrne R.W. 1994). Based on what was said above, the reason would be that selection on fitness alone is not sufficient. First it needs an accidental investment caused by mutual selection on social competence. Once the development of a bigger brain has started in this manner, it can eventually also result in an increasing fitness.

Homo sapiens and the new phase of evolution

Now, let's compare the assumptions made about the entry of a species into the unrestricted niche phase of evolution with what is known about the rise of *Homo sapiens* (for the quoted details of the human evolution see again Leakey R. 1994 or Tattersall I. 1998).

There is no doubt that *Homo sapiens* started a new phase of complexity increase that resulted in the development of increasingly complex human societies. But does our evolutionary history also match with the unrestricted niche model? The answer is yes, it does. Until about 30-40,000 years ago, the genus *Homo* was represented by different species living at the same time. Our own species emerged very probably 150-200,000 years ago in Africa. At that time *Homo sapiens* represented not more than just a new twig on the homo branch, which had to share the planet with its cousins such as *Homo erectus* and *Homo neanderthalensis*. The coexistence of different *Homo* forms is in line with the law-like process of complexity increase and corresponds to the situation of a respective diversity within established ecosystems (compare flowchart 1 – step 1).

Suddenly, however, about 70,000 years ago something happened. Our ancestors started to leave their African homeland and to conquer all possible climate zones and habitats around the world. Only about 30-40,000 years later, which is very fast in evolutionary terms, *Homo sapiens* was well presented all over the world – with the exception of the Americas, which followed for geographical reasons a bit later. In parallel other *Homo* species, such as the Neanderthals, disappeared (for Neanderthals see in particular Patou-Mathis M. 2006). Obviously this corresponds well to the predicted scenario for the entry of a highly intelligent species into the unrestricted niche phase. But what was the reason for this sudden development? Following the law-like process of complexity increase, this step would need a large-scale environmental catastrophe disrupting the established ecosystems. Well, it seems that the exodus out of Africa coincides indeed with dramatic worldwide environmental changes. About 75,000 years ago the earth experienced the so-called Toba catastrophe. The explosion of the super-volcano Toba, in what is now Indonesia, was gigantic and much more dramatic than all known volcanic catastrophes in more recent times. The consequences for the global climate must have been devastating for many thousands of years. As an example, the temperature of the oceans went down on average by about 5° Celsius in the period following the Toba event, which gives an idea of the dramatic worldwide changes (Burroughs W.J. 2005).

The migration of *Homo sapiens* out of Africa resulted in the fast adaptation to all possible climate zones and habitats. This is very remarkable for a species that evolved under the warm, tropical sun of Africa. For instance, modern humans started to infiltrate ice-age Europe about 40,000 years ago and some thousand years later the Neanderthals became extinct. In other words, the locals who were adapted to the cold and unpleasant conditions ruling in Europe at that time were ousted by people coming from tropical Africa. Obviously this was only possible because of strong cultural and cognitive accomplishments (compare Mellars P. 2006). One of these accomplishments could have been trade and long-distance relations, which was very probably much more developed by *Homo sapiens*, than by the much less interconnected Neanderthals, as far as we know (e.g. Tattersall I. 1998). Again both the fast migration into all possible climate zones and the disappearance of other *Homo* species fit well with the unrestricted niche model.

But not only within the *Homo* genus had the entry of *Homo sapiens* into the unrestricted niche phase left its traces. Many species of big game, such as the well-known example of the mammoths, as well as several top predators, disappeared following the arrival of our prehistoric ancestors (see e.g. Burroughs W.J. 2005). Again, this corresponds to what can be expected from the unrestricted niche model.

In summary it can be said that what is known about the rise of *Homo sapiens* matches well with the predictions made about the entry of a highly intelligent species into the unrestricted niche phase of evolution.

New forms of mutual selection

Another striking aspect in the evolution of *Homo sapiens* is the exceptional importance and dramatic intensification of mutual selection. It provides additional support for a key message of this book, namely that complexity increase is accompanied by an intensification of mutual selection.

Like no other species, we are formed and conditioned by our social environment (see e.g. Tattersall I. 1998). It strongly influences our social competence as well as our behaviour and skills. Such an unprecedented high degree of social interactions with other members of the own species corresponds to what was predicted regarding a necessary intensification of mutual selection with ever more complex forms. Humans own kinds of reproductive interactions which are not present in any other species. A good example is the role of spoken language, which can be seen as such a human-specific interaction with influence on the individual multiplication. It causes an entirely new type of interaction community which is formed by all those individuals speaking the same language. The fact that language is much more than a simple communication tool becomes visible in its impact on the individual multiplication success. If I do not speak the same language as my social environment, I am normally isolated, with no or a very limited prospect of finding a partner. On the other hand, if strangers discover that they speak the same language they have suddenly something very central in common. Imagine for example a group of prehistoric hunters who encounter unknown folks in the wilderness. It is obvious that they will react differently if they speak the same language or not. From the *Interaction Theory* point of view, language must therefore be seen as a central part of our individual human interaction competence, which causes a mutual selection. In other words, language was and still is, without any doubt, a key criterion for social selection and recognition (for the evolution of language see e.g. Fitch W. T. 2010 and for its social implications e.g. Taylor T. 1996 or Miller G. 2001).

With regard to complexity increase, probably the most important consequence of language is that it overcomes the need to know an interaction partner personally.

This fact was very probably central for the success of our species. In the case of other, highly social forms such as chimpanzees or dolphins, the members of a social interaction community know each other and the arrival of unknown individuals does normally cause stress (Lee P.C.. 1994). This close link between social interaction and the need to know each other personally can change with the development of language. Now it is possible that strangers also belong to the same community, if they speak the same language. In addition, language allows the identification of social status. Very similar in this respect are tools and personal ornamentation. As an example, by the use of a certain type of artful arrowhead, a hunter can not only communicate to which group or tribe he belongs, but also his social status (e.g. Mellars P. 2002). The same can be said about personal ornamentation as well as the emergence of art in general (see Taylor, 1996). This means that personal features such as clothing, ornaments, weapons or artful tools correspond to visible signs which document the belonging to a certain tribe or social status.

In summary it can therefore be said that humans are indeed exposed to an unprecedented intensification of mutual selection, which is mainly caused by new kinds of interactions. So thanks to language it was no longer as essential to know each other personally. Consequently, the human interaction communities became much larger and thus the mutual selection more sophisticated, because the individuals had to develop new ways of signalling the individual status and interaction competence also to those who did not know them personally. Once this state is reached, it can be assumed that it represented the basis for trade and strong long-distance relations between geographically dispersed groups. As mentioned before, here lies very probably a decisive advantage which allowed *Homo sapiens* to conquer the world.

The archaeological record supports this view, for instance with regard to the sudden appearance of rich personal ornaments and artful tools. Starting about 50,000 years ago a sudden increase in the variety and sophistication of tools, such as arrowheads and personal ornaments, became detectable. This phenomenon is often called the Palaeolithic Cultural Revolution, which is in particular clearly visible in Europe. The earliest known signs of this revolution appeared in Africa about 70,000 years ago (Mellars P. 2002 or Tattersall I. 1998). This would fit the above-mentioned role of the dramatic Toba event which occurred just before and forced Homo sapiens to try his luck also outside Africa.

At this point, I come to the end of the discussion about the destiny of generation sequences travelling through deep time. As already stated at the beginning of this book, I'm deeply convinced that *Interaction Theory* provides meaningful insights into the biological complexity increase and therefore can replace the conventional understanding of a solely fitness-driven evolution. Here I can only leave it to the reader to decide whether she or he shares this opinion.

Before ending this book in the last chapter with some personal thoughts about the possible consequences of *Interaction Theory* beyond biology, I try in the following to give a new answer to the central question in natural science: "what is life?"

Part 3: What is life?

At the beginning of this book, I mentioned that a satisfactory answer to the fundamental question in biology "what is life?" is still pending. On the one hand, the understanding of the biochemical and molecular basis of life is growing steadily, as for example by the decoding of the entire genomic DNA sequences in an increasing number of species. On the other hand however, it is still difficult to exactly pin down the decisive difference between a living being and a dead object (see e.g. Murphy M. P. & O'Neill L.A.J. 1995). Here, *Interaction Theory* can provide a radically new answer that is based on the insights about generation sequences and the law-like process of complexity increase. Thus, the answer proposed here to the question "what is life?" is the consequence of the following insights: life manifests itself in the occurrence of generation sequences proceeding over deep time. The sequences are caused by the sustainable multiplication with variations of forms that multiply either solitarily or as part of interaction communities, depending on the respective kind of environmental conditions necessary for the multiplication. The generation sequences are exposed to the law-like process of complexity increase that produces over time in some sequences increasingly complex forms characterized by a progressively rising probability of multiplication.

This description of life is centred on generation sequences and therefore not very practical to decide whether a particular structure or object which might be found one day on another planet deserves the attribute life. For this purpose, a definition of living forms is helpful that is not dependent on the observation whether the respective structure or object is part of an ongoing generation sequence or not. From the *Interaction Theory* point of view such a definition is possible, since it was said that the generation sequences of life can only get started with spontaneously emerging molecular interaction communities. As outlined, these communities would consist of different types of relatively simple, molecular forms, which need each other in order to become multiplied. On earth, the respective molecules would have been the precursors of DNA, RNA, protein and biological membranes. In the first phase of complexity increase, the basic cellular metabolism would then have developed around this reproductive interaction. This means that the development of the basic cellular metabolism is driven by the need to ensure the ongoing multiplication of the interacting molecular forms via the transformation of environmental resources into the required energy and building blocks. This results in the following definition:

- A living being is a structure or form that emerges from a multiplication under variation and represents the outcome of a generation sequence. Biochemically it is characterized by an underlying reproductive interaction of different molecular forms and the capability to transform environmental resources into the necessary chemical energy and building blocks, for the multiplication of this molecular interaction community.

Hence if one day a structure or object is found on another planet it can be biochemically analysed as to whether it fulfils the given criteria. If yes, it deserves the attribute alive, if not it is dead matter. For example, on earth life is based on the reproductive interaction of DNA, RNA and protein via the biochemical processes of transcription and translation. The metabolism transforms environmental resources into the energy and substances that are necessary to sustain the transcription and translation process. In contrast to traditional definitions, viruses would thus also correspond to living forms, because they fulfil the above demands by using their hosts as relevant environmental resource.

The *sine qua non* of living forms is therefore an underlying reproductive interaction of different molecules. Hence, a living form without such a reproductive interaction would not be possible, and this dependence of life on corresponding interactions would be irrespective of what the interacting molecules exactly look like. If it is further taken into account that the generation sequences of life need a specific starting point, in the form of the spontaneously emerging molecular interaction communities, the definition of life as such can be phrased as follows:

- Life is the incessantly ongoing multiplication of reproductively interacting molecular forms, thus with a mutual dependence of their respective multiplication. For probability reasons, these forms consist of a limited number of building blocks, such as polymer-like molecules or lipid bilayers. On earth these are DNA, RNA, proteins and biological membranes. Life starts with spontaneously emerging molecular interaction communities of relatively simple corresponding precursor forms. Due to exhaustible resources and the need for sustainability, the multiplication of these very first multiplying forms becomes subjected to the law-like process of complexity increase. This corresponds to the starting point of the generation sequences of life, which can continue over deep time due to the stepwise emergence of multiplying forms with a higher probability of multiplication and thereby ensuring the ongoing multiplication of the underlying molecular interaction community.

With the above definition, life becomes an inherent quality of matter. In other words, whenever and wherever molecules organize spontaneously as reproductive interaction communities, they become the subject of the law-like process of complexity increase, and the evolutionary auto-complexification can get started. This need of corresponding molecular interaction communities for the start of generation sequences can explain why life depends on water and certain temperature conditions. The fundamental molecular interaction is based on non-covalent and reversible bindings between the different interacting molecules. And those kinds of bonds, such as hydrogen bonds and ionic bonds, depend on water and certain temperatures (for the physicochemical conditions at the beginning of life see e.g. De Duve C. 1995 or Davies P. 1999).

Now, if life is an inherent quality of matter and the evolution driven by the law-like process of complexity increase, it can have consequences for our self-image as humans. The reason is that the process of complexity increase towards a rising probability of multiplication would have followed the different, illustrated phases. As a result, not only life as such would be an inherent quality of matter, but also the potential for an intelligent and conscious being. Therefore, if planets are frequent in the universe and a part of them provide the conditions for the start of generation sequences, it can only mean that we are not the only intelligent living beings that exist[11].

---

[11] Here I should probably also include those intelligent beings in other parts of the universe that already existed or will still exist.

Part 4: The consequences of *Interaction Theory* beyond biology

In the last part of this book I want to share some personal thoughts about the possible insights that *Interaction Theory* can provide about the development of human society.

The central message of this book is that the evolution of life follows a law-like process of complexity increase to biological forms with a progressively rising probability of multiplication. I'm convinced that the fundamental principles of this law-like process are also applicable beyond biological evolution to any kind of evolutionary auto-complexification. In other words, the key insights of *Interaction Theory* can also be seen as relevant for the evolution of global human society. As a consequence, *Home sapiens* historical development from the first prehistoric communities to modern, highly complex societies would correspond to the seamless continuation of the biological complexity increase. In the following I will try to convince the reader of the correctness and importance of this conclusion. The reason why I consider this as important arises from the conviction that a fundamental paradigm shift in the sense of Thomas Kuhn (see Kuhn T. S. 1962) is urgently necessary in our collective consciousness: we need to overcome the prevailing misconception that the current global growth contest results in progress. Instead we need to replace it with an understanding that any form of a global growth contest cannot be sustainable and that true progress, in the sense of a further and more balanced rise in the welfare for all of humanity, depends on new forms of mutual interaction.

The relevance of *Interaction Theory* beyond biology

From *Interaction Theory* it can be deduced that if a form, structure or object, with the capacity to multiply with variation, is exclusively exposed to a selection on fitness or fitness-like criteria, it will inevitably get trapped in a general growth contest. This means that the form, structure or object transforms into what I described as a multiplication bomb. It will entirely become adapted to react with fast, self-accelerating growth, whenever the circumstances allow it, with all the consequences for the environment. As already said many times, this can only be sustainable for forms such as prokaryotes. They own niches which can tolerate the general growth contest, because the resources or conditions can regenerate after periods of exponential growth. In a situation however, where the conditions can become irreversibly exhausted or destroyed by self-accelerating growth, the sole selection on fitness or fitness-like criteria is dangerous. The resulting general growth contest bears the risk of destroying over time the conditions that are necessary for the multiplication. For this reason nature depends on reproductive interactions, such as sexual reproduction. And in the end it is this need for interaction and the associated mutual selection which causes progress, in the form of a rising complexity with a progressively increasing importance of the individual. *Homo sapiens* is the outcome of this biological auto-complexification, and I can see no reason why the causal link between complexity increase and sustainability by mutual selection should not be valid beyond biology.

Biological evolution and the development of our human society show fundamental parallels. Both depend on systems, ecological or socio-economic, in which the respective actors are connected via symbiotic and competitive relations. In both cases, the system evolves over time and shows an auto-complexification towards a rising probability of multiplication. For example, the probability of multiplication has reached in the modern, highly developed societies an all-time high, which is visible in a strongly decreasing birth rate on the one hand and a dramatically decreasing child mortality and greater than ever life expectancy on the other. Furthermore, the probability of multiplication rises thanks to the introduction of new techniques such as *in vitro* fertilization or uterine transplantation. It allows otherwise childless couples to procreate and to continue their respective generation sequences, which would otherwise end. From the *Interaction Theory* point of view this is exactly what can be expected, if the development of the human society continues the evolutionary auto-complexification process towards a rising probability of multiplication. The same can be said regarding the increasing importance of the individual that is visible through the introduction and consequent application of individual human rights by an increasing number of countries. Again this is consistent with what has happened during biological complexity increase and corresponds to a continuation of the identified development from quantity to quality.

An indispensable driver for biological evolution as well as for the development of our society is the competition between the individual actors. In biology these are the individual multiplying forms as well as interaction communities, such as an ant colony. In human societies, the competition takes place between individuals as well as organisations, such as companies or national states[12]. Essential in this regard is that competition can occur in two ways. It can take place either via a general growth contests, resulting in a rollercoaster-like growth pattern, with the risk of exhausting the resources or conditions. Or alternatively, it can occur in a more balanced, less rollercoaster-like manner, thanks to interactions causing a

---

[12] From the *Interaction Theory* point of view, human organizations such as companies and states can be seen as interaction communities.

mutual selection that can build a barrier against exponential growth. In my view, precisely here lies a fundamental problem with our current social and economic situation: we are in the grip of an unhealthy general growth contest, bearing the risk of an irreversible exhaustion and destruction of our planet. As in the case of biological systems, we also have an urgent need for new, adequate kinds of interaction with new forms of mutual selection that can make further social and economic development more balanced and sustainable.

The obsession with growth

In the past few decades societies around the world became obsessed with economic growth and the restless pursuit of profit. In this regard it is striking that the desire for growth and profit plays a role in modern economy that is comparable to fitness for biological evolution. For example, from the *Interaction Theory* point of view, an exclusive selection on fitness or fitness-like criteria results in self-accelerating growth. And indeed, also in economy it is well known that free and unregulated competition for profit drives overall economic growth. And as the exclusive selection on fitness in nature, the exclusive selection on profit causes a rollercoaster-like pattern, which is in particular clearly visible in fast-moving markets, such as the global stock and financial market. As an example, the last global financial crisis demonstrated clearly that a fierce worldwide competition for ever-increasing profits causes strong growth that is eventually followed by a sudden general crash. This rollercoaster between growth and crash that results from a general growth contest had negative consequences not only for the financial markets around the world, but for the stability of the whole economic system. In this particular case it was in the end the taxpayer who had to correct the situation so that the financial system did not become completely ruined. In other economic sectors, the consequences of a general growth contest are much less reparable by the taxpayer's money. Here I mean sectors such as the auto or airline industry, in which an ongoing overall growth causes an accelerated consumption of non-regenerative resources and the destabilisation of the world climate by CO_2 emissions. Hence, if the global economy continues with a ruthless general growth contest, it will not only endanger economic and social stability, it also endangers the ecological balance of the whole planet earth (for the consequences of economic growth see e.g. Ekins P. 2000).

It is without any doubt that the further development of modern human societies depends on the competition for economic success, exactly as biological evolution depends on the competition for multiplication success. On the other hand, competition means normally that only a few win and the majority loses, which represents a fact that is difficult to accept for a democratic society. From the socio-

166

political point of view, the obsession with general growth by the governments around the world is therefore partly understandable. In a society with general economic growth, most individuals have a direct or indirect economic benefit, even if normally the gain for the masses is rather modest, while a small group profits the most (see e.g. OECD Wealth Distribution Database). As a consequence, in democratic, capitalistic societies general economic growth is in general perceived as good for the social stability. However, this stability is obviously an illusion, because modern economies are in permanent danger of falling from growth into recession. From the *Interaction Theory* point of view this is not a surprise, because a general growth contest provokes inevitably rollercoaster-like fluctuations. And as said, since this general growth contest depends on exhaustible resources or conditions, it cannot be sustainable either.

The consequences of the current obsession with growth are, however, not only visible on the level of the global financial and economic markets; they also influence our personal daily life. By this I mean the general development towards a situation in which more and more of us experience the feeling that nothing in life can stand still anymore and the wheel, so to speak, is turning faster and faster. For many it goes along with the perception of a steadily increasing pressure for personal fitness in the widest sense. For example, as individuals we are not only exposed to a permanent demand to be fit, physically as well as mentally, we also have constantly to adapt and change, otherwise we may not be able to keep the pace. In the end we come more and more into a situation comparable to our gut bacteria. They too, experience the constant pressure not to fall back behind their peers. Obviously we do not compete for the same things as bacteria; we compete for jobs, money and social recognition. But irrespective of such differences, what counts is that in our fitness- and growth-obsessed societies we find ourselves more and more in the previously mentioned red-queen situation. Not much different than our gut bacteria, we need to run faster and faster just to stay in the same place (see the previously mentioned Red Queen effect).

In summary, governments worldwide entrust the future and well-being of their nations to the belief that only a strong persistent growth of their national economies can provide social stability and cause progress. As a result, growth as such became more and more a goal in itself, ignoring all too often the distinction between unsustainable and sustainable growth. In my personal view, it is rather scary that in the age of a global economy, with its ruthless competition for money and jobs, this naive belief in growth carries almost religious qualities.

New forms of interaction in human evolution

A central finding of *Interaction Theory* is that complexity increase towards a mounting probability of multiplication is accompanied by the emergence of new kinds of interaction resulting in a progressive intensification of mutual selection. As said, in our modern, highly developed societies the development towards a rising probability of multiplication is documented by a decreasing birth rate on the one hand and a greater than ever life expectancy and importance of the individual on the other. At the same time, we risk becoming victims of an unsustainable growth contest, which provokes the question why? In view of the above central finding, the answer would be that it is the necessary intensification of mutual selection which would currently not be adequate. In other words, if the insights of *Interaction Theory* are applied to the further development of the global human society it means that we need urgently new kinds of interaction that allow additional mutual selection. The resulting intensification of mutual selection would be the means to overcome the current global growth contest, which endangers not only the environment, but also the social and economic balance in our increasingly complex societies. The billions of years of biological evolution show that this is the only way that can lead to true progress, in the form of a further rising probability of multiplication, also for those parts of the world where the birth rate is still too high and overall life expectancy still low.

Our own species provides impressive examples that complexity increase goes along with the emergence of new kinds of interactions, with new kinds of mutual selection. Let's take love, which is a particular and primary form of human interaction with a strong and direct impact on reproduction. As human beings we do normally not marry each other based on fitness criteria; we need to fall in love. From the *Interaction Theory* perspective it is thereby not essential why exactly we fall in love. It is rather essential that it causes an additional mutual selection, which is independent of selection on fitness. In other words, love ensures that we do not always and exclusively choose "the fittest" as partner. For this reason, love should be understood as an invention by nature that builds an extra barrier against an unsustainable growth contest. The contest would occur if human reproduction were only driven by the desire to have more progenies than others. The connection of human reproduction with the need to fall in love can therefore reduce the risk of unsustainable population growth.

A further example that the evolution of modern societies was accompanied by the emergence of new kinds of interaction is the following. The most developed countries, in which the economic, social and cultural complexity of modern life is the highest, are mainly democracies, while countries in the underdeveloped parts of the world are often autocratic. And compared to autocratic regimes, the democratic system causes obviously a new kind of interaction with additional mutual selection. By this I mean the interaction between those governing a country and those electing the government. This, let's call it "democratic interaction" has obviously a strong impact on the general living conditions in the respective society and thus influences reproduction-relevant aspects, such as family planning and number of children.

Towards a Consumer-Guided Economy

Now, what can a new form of interaction in our society look like, so that it can build an additional barrier against unsustainable growth? An obvious idea is that it is no longer sufficient to vote on average every four or five years for a national government with limited power and influence on global developments. As a result of the dramatically rising complexity in our globalized world, we need new ways in which to act against the danger of an unsustainable global growth contest. This in particular in highly cross-linked sectors, such as the economic and financial markets.

Let's take for example big, multinational businesses. For them it would not be enough to be only exposed to a fitness-like selection, which is caused by changing market situations and the pressure from stock markets to maximise profits. To make their economic activities more sustainable, socially and economically, it needs an additional, permanent and strong selection by consumers. In other words, we need a new kind of interaction between consumers and companies that influences the further economic development. This is in particular necessary, because only consumers can liberate the economy from the current dictatorship of the shareholder value. Hence, we need what can be described as a Consumer-Guided Economy. In order not to be misunderstood, by this I do not mean the rather banal choices and decisions that currently drive general consumer behaviour. Those are often simply the result of targeted manipulations by pushy marketing departments. In contrast, what I mean by Consumer-Guided Economy is that consumer behaviour will increasingly be the result of a conscious and responsible decision-making process. Here, the large majority of consumers will see products and services they buy or consume no longer solely under aspects such as price, convenience or fashion. Rather, they will also intentionally use the purchasing act to influence the social and environmental impact of a chosen product or service category. This does not necessarily require that all consumers perform each and every purchase in this way. It would be already a great step forward if they participate regularly in well prepared and coordinated campaigns which target a particular issue or sector. An example would be a campaign that avoids products from companies which distinguish themselves by unjustified high profits or exaggerated management compensations. Another example would be a campaign in which consumers only buy clothing from companies which do no longer exploit the social and economic situation of textile workers.

For this purpose, consumers need to organize themselves into new kind of interaction communities beyond national borders. The main activities of these communities would be the following. They have to constantly evaluate products and services on the basis of defined and agreed sustainability criteria. Their members as well as the general public will be periodically informed about the respective results and, where it is seen as necessary, they will organise coordinated and targeted campaigns. Beside ecological aspects, such as CO_2 emissions, this would in particular include economic criteria. As an example, let's take the fight against exaggerated company profits. The main issue is in this regard that the current dictatorship of shareholder value causes an excessive redistribution of consumer spending into shareholder profit. In the end this is unsustainable, because the shareholders are mostly financial investors who do often not spend the earned money in the real economy. They rather re-invest it in the financial markets, with the danger of inflating the financial system and causing new bubbles (see e.g. Vogel H. L. 2010). In other words, an economy that is only driven by shareholder value causes an unsustainable redistribution of money from the real economy into financial speculations. Consequently, by buying products or services from a company that is solely focused on shareholder value, consumers unconsciously support this unsustainable trend. However, if a well-organized consumer campaign exerts pressure on companies to avoid exaggerated profits, it would be possible that a much bigger part of the company earnings will flow back into the real economy. In other words, instead of constantly delivering a maximum to shareholders, the companies could invest more into new company equipment or pay higher salaries to their employees. In the end it will be beneficial for the economy and its stability as a whole. No economic or political actor other than consumers would be able to achieve this (for shareholder value compare e.g. Stout L. A. 2013).

Obviously, comparable kinds of consumer activities already exist. In general however, they occur on a very limited scale and are only supported by a relative small number of people, compared to the mass of consumers. In a Consumer-Guided Economy, however, it would be on a much larger scale that respective targeted campaigns take place. For this, it is essential that the general public develops the awareness that it is a central part of civic duty to participate actively in corresponding movements. For those who believe that this is unrealistic, just consider the history of democracy. A few hundred years ago nobody could imagine that a majority of people would be willing and able to actively participate in a fully democratic society. And exactly as this has happened, I'm deeply convinced that it is only a matter of time until a Consumer-Guided Economy will drive businesses in all sectors into new, much more sustainable ways of competition. This means that the time of focusing solely on profit maximisation will be over. Now, all businesses have to convince their customers not only that their products or services are good, but also that they are socially and ecologically responsible.

In the same way a strong Consumer-Guided Economy can also act against the rollercoaster effect mentioned, and help to make the market dynamic more balanced. The above-mentioned fight against exaggerated profits would be an example of how consumer selection can contribute to a more balanced market dynamic. Another example would be the following. At the moment a national economy slides into recession, most companies react with measures to defend their profit targets. This means that they try to ensure a high profit by reducing their spending. It is obvious that when companies also spent less money it will accelerate and deepen a recession. Again a coordinated, well organized consumer movement can act against such company behaviour, by supporting businesses which accept under these circumstances a profit reduction and contribute in this manner to more economic stability.

How can consumers organize in a new way, so that they are able to participate in activities or campaigns, with the goal of motivating businesses or sectors to be socially and ecologically more sustainable? Here the modern, mobile internet devices with social media and instant messaging technology can play a decisive role. They provide unprecedented possibilities for organized and coordinated activities. Thus, when a significant proportion of consumers is ready to be part of such digitally connected interaction communities, it becomes possible to fight unsustainable and socially destructive corporate behaviour with powerful, internationally coordinated campaigns. The consequences for planet earth and our all future would be fantastic. No government or governmental organisation can achieve the same. The Consumer-Guided Economy can change the world in a way that is comparable to what the emergence of political democracy has accomplished.

At this point I want to end the discussion about the consequences of *Interaction Theory* beyond biological evolution; it can by itself provide the subject for a book. With this, I am also at the end of the presentation of *Interaction Theory* about the extraordinary evolutionary auto-complexification of matter, from spontaneously emerging molecular interaction communities to modern digital societies.

Flowchart 1

The law-like process of complexity increase

1. Generation sequences can only continue in the 2 fundamental types of niche that either need a reproductive interaction or not; the niche adaptations lead to a diversity of forms and ecosystems, in which the available niches are largely occupied over time.

   ↓

2. The multiplying forms carry qualities that are either fitness related and a result of selection on fitness, or are originally non-fitness related since they are caused by mutual selection resulting from interaction or symbiosis, depending on the respective type of niche.

   ↓

3. Qualities caused originally by mutual selection or symbiosis can start influencing the fitness and therefore become also an object for selection on fitness; if the now dual selection has a positive impact on the environmental adaptation, it can result in a new, distinct phylogenetic group.*

   ↓

4. Over time key innovations also emerge in this way, if a dually selected quality makes it possible to overcome the dependence on a hitherto essential niche factor, by strengthening a weak-point in the lifecycle of the respective form at which most of its generation sequences get stopped.

   ↓

5. As a result, the key innovation enables adaptation to correspondingly special niche situations, so that alongside the forms without this key innovation, which dominate the established ecosystems, a new phylogenetic genus emerges.**

   ↓

6. Over time a big catastrophic change or event destroys the established ecosystems; thanks to their strengthened lifecycle, the key innovation-owning forms survive better than the hitherto dominant forms and can prevail in the new ecosystems that develop after the disaster.

   ↓

7. Consequently, the key innovation-owning forms can now also occupy ordinary niche situations previously occupied by the formerly dominant forms; due to the strengthened lifecycle this requires, however, a development from quantity to quality in order to be sustainable.

   ↓

8. As a result, the now dominant group of key innovation-owning forms shows on average a higher probability of multiplication than the group that dominated the ecosystems before the large-scale catastrophic change or event: hence, a complexity increase has occurred.

   ↓

9. Over time the available niches in the ecosystems become largely occupied and the process starts again at the beginning.

The repetitive succession of rounds results in a clustered organisation of the biological diversity, in which the more recent phylogenetic groups have a higher average probability of multiplication and therefore a rising complexity.

*groups such as dog- or cat- like carnivores
** such as mammals during the reign of the dinosaurs

Annex

<u>Brief summary of the main terms and insights of *Interaction Theory*</u>

Generation sequences and multiplying forms

Interaction Theory is based on the insight that the key characteristic of life is the emergence of generation sequences continuing uninterrupted through deep time in a world with limited and exhaustible resources. For this reason a key demand is that the reproduction of the biological forms, which causes and drives the generation sequences, needs to be sustainable. The reproduction of biological forms corresponds, however, to a multiplication in the sense of 'making more'. Thus, biological forms are potential multiplication bombs, ready, whenever possible, to exhaust by exponential growth the environmental conditions on which they depend. The solution to the resulting conflict between sustainability and multiplication lies in the nature of niches and the role of reproductive interactions.

The two fundamental types of niche

Environmental conditions can only serve as a niche for multiplying forms if they provide sufficient stability and reliability for the respective generation sequences so that they can continue over time. Here, two fundamental types of niche can be differentiated. The first type corresponds to niches that can tolerate exponential growth, because of resources and conditions that are able to regenerate after their exhaustion or destruction by overconsumption. These niches can harbour the generation sequences of solitary multiplying forms, such as prokaryotes. The second type comprises niches that are at risk of becoming irreversibly exhausted or destroyed by exponential growth. These niches can therefore sustain only the generation sequences of reproductive interaction communities in which the individual multiplication success is under the influence of mutual selection. Sexual reproduction corresponds to such a reproductive interaction. Consequently, for generation sequences two fundamental types of niche exist, depending on whether their environmental conditions tolerate exponential growth or not.

Fitness, multiplication potential and selection on fitness

The fitness of a biological form dictates its multiplication potential, corresponding to the in principle possible number of progenies (N/t) under the given environmental conditions (EC). For solitary multiplying forms adapted to exponential growth tolerating niches, the individual multiplication success (Msuc) corresponds therefore to the respective multiplication potential – according to the equation $Msuc = N / t$ (EC). For forms with reproductive interactions, such as sexual reproduction, the term fitness refers also to the multiplication potential due to the respective, individual environmental adaptation. However, this does not correspond to the multiplication success, since in this case also the respective individual competence to gain a reproduction partner is essential. In both cases, selection on fitness results therefore from differences in the multiplication potential of biological forms which compete for the same environmental or niche conditions. Hence selection on fitness favours the variants of a form with a relatively higher multiplication potential due to a better adaptation to the given environmental conditions. As a result, an exclusive selection on fitness provokes a general growth contest in which all forms multiply and those with the highest multiplication potential eventually prevail.

Reproductive interactions and mutual selection
A reproductive interaction describes a situation in which a biological form cannot multiply alone, but requires a partner, as is the case with sexually reproducing organisms. Reproductive interactions are mandatory for the sustainable occupation of niches at risk of becoming irreversibly exhausted or destroyed by growth. Their impact on growth results from the fact that within a particular interaction community the respective forms must compete for interaction partners in order to multiply. This causes a mutual selection on interaction competence, which, however, must be independent of selection on fitness. If this is the case, the reproductive interaction reduces the risk of a general growth contest, because the multiplication success (Msuc) of an individual variant no longer depends on their multiplication potential only (N/t), but rather also on their particular interaction competence (IC) – according to the equation $Msuc = IC \times N / t \, (EC)$, , with IC varying between 1 and 0.

Dual selection
Dual selection stands for the simultaneous effect of mutual selection and selection on fitness. It therefore includes both the interaction competence and the fitness of a multiplying form. Hence, dual selection can reduce the risk of exponential growth, compared to an exclusive selection on fitness. This means that an individual with a high multiplication potential but a low interaction competence may only have a low or no multiplication success. In this way, dual selection can build a barrier against a selection on fitness-driven general growth contest and thus ensures a sustainable balance between multiplication and environment. For this it is however mandatory that both types of selection are independent, which explains for example why genetic recombination is an indispensable part of sexual reproduction.

Sexual reproduction and species

Sexual reproduction corresponds to a main form of reproductive interaction during biological evolution, and sexually reproducing species must be seen as interaction communities which need the mutual, here sexual selection, for the sustainable occupation of their niches. The genetic recombination as part of sexual reproduction ensures that qualities relevant for interaction competence and those relevant for fitness can stay independent on the genetic level. In this way the dual selection on both, sexual attractiveness and fitness, can reduce the risk of unsustainable growth.

Symbiotic interaction

A symbiotic interaction is given, if different forms have a positive mutual influence on their respective multiplication but could still reproduce individually, this at least in principle. Symbiotic interactions can also cause a kind of dual selection and are therefore also relevant for complexity increase. While they can also occur in exponential growth-tolerating niches, reproductive interactions occur only in niches at risk of becoming irreversibly exhausted or destroyed by exponential growth.

Probability of multiplication

The more complex a biological form is, the longer it needs for its reproduction and therefore multiplication. The frequency of branching of generation sequences of more complex forms is therefore lower. Compared to faster multiplying forms with lower complexity, the generation sequences of more complex forms must therefore be less frequently stopped on average, to stay in a sustainable balance with the environment. This can be expressed as follows: complexity increase goes into the direction of a mounting probability of multiplication. This term stands for the average likelihood by which an individual generation sequence of a particular form can proceed over a longer, multiple generations encompassing a period of time, without being stopped. The probability of multiplication can therefore be determined via the ratio of non-stopped generation sequences, ranging over the respective time period, with those which theoretically would have been possible if none of the sequences had been stopped. An important consequence hereof is that in a world with limited and exhaustible resources, the probability of multiplication cannot significantly increase in growth-tolerating niches. For this reason, complexity increase occurs mainly in growth-sensitive niches. To be sustainable in these niches, a mounting probability of multiplication needs a development from quantity to quality, in the sense that the average multiplication potential, thus the in principle possible number of offspring per time, goes down, while in parallel the investment into the single offspring increases. More complex forms are therefore characterized by a relative higher investment into the single offspring, so that the individual generation sequence has an on average higher probability to continue over time, compared to less complex forms. This has consequences for the niche situation, in the sense that more complex forms need environmental conditions with on average relative higher stability or reliability for the individual generation sequence.

Weak points in the lifecycle

For a rising probability of multiplication via a development from quantity to quality, it needs a strengthening of a decisive weak point in the lifecycle, where normally most generation sequences of the respective form get stopped. For higher organisms, such a decisive weak point is for instance the vulnerable early phase of life. A corresponding strengthening can occur in two principal ways. Either it is connected to a particular niche situation and driven by a selection on fitness. Or it is not caused by a particular niche situation and is dependent on a dual selection. The second way would cause the clustered organisation of the biological diversity along the phylogenetic tree of life.

Key innovations and taxonomic clusters

Key innovations are different from other adaptive qualities, because they cannot emerge via an exclusive selection on fitness. They rather need the consecutive action of mutual selection and selection on fitness, thus dual selection. The emergence of key innovations drives the biological complexity increase and is responsible for the emergence of new phylogenetic clusters in evolution. Key innovations are therefore those qualities that make a plant an angiosperm or an animal a mammal. The dual selection necessary for the emergence of key innovations results either from reproductive or symbiotic interactions, depending on the type of niche. With regard to reproductive interactions, such as sexual reproduction, this means that they are not only essential for the sustainability of generation sequences in niches not tolerating exponential growth, but deliver also the basis for complexity increase during evolution. Alternatively, solitary multiplying forms in exponential growth-tolerating niches need symbiotic interactions for the emergence of key innovations via dual selection.

The benefit of key innovations

The reason why key innovations require dual selection is that they are good for the generation sequences, but less for the individual form in the here and now. Key innovations act via the strengthening of a decisive weak point in the lifecycle of the corresponding form. This is done by overcoming or at least significantly reducing the multiplication's dependency on a hitherto essential niche factor or condition that has a limited reliability for the maintaining of generation sequences. This has consequences for the niche adaptation. As immediate effect, the multiplication can now occur under different environmental conditions, which are not possible for otherwise comparable forms, but without the key innovation. As a result, the key innovation-owning form can adapt to corresponding, particular niche situations. This becomes visible in a new, but limited phylogenetic cluster within an established ecosystem, dominated by forms without this key innovation. And in addition, because the key innovation strengthens a weak point in the individual life cycle, it can also provide the following, additional effects: In ordinary or general niches, for which the key innovation is not mandatory, it makes the niche conditions more reliable or stable for the individual generation sequence, so that less of them become stopped per time. As a consequence, the sustainable adaptation to this kind of non-specialist niches requires a development from quantity to quality and herewith a rising probability of multiplication.

The role of large scale environmental catastrophes
In established ecosystems newly emerging forms with key innovations can normally not oust the dominant forms from their niches. Over deep time however, catastrophic large-scale events or changes occur in the environment and destroy established ecosystems. With their more robust lifecycle, because of the strengthening of a respective weak point, the key innovation-owning forms are more likely to survive these catastrophes than comparable forms. As a result the key innovation-owning forms can become dominant in the newly developing ecosystems and now also adapt to those niche situations which were previously occupied by the formerly dominant forms. To be sustainable, the respective niche adaptation requires a development from quantity to quality in all those cases where the key innovation provides more reliable or stable conditions for the individual generation sequence, as it was the case for the dominant forms before the catastrophe.

Biological complexity
The biological complexity determines the average likelihood by which the individual generation sequence of the corresponding form can continue over time. The most complex organism at a given time is that of which the individual generation sequence owns the highest corresponding likelihood. Thus, the complexity increase during evolution goes into the direction of a rising probability of multiplication and is the product of a stepwise accumulation of key innovations via a law-like process. This means, the complexity of a living being is the manifestation of the totality of its key innovations that have been acquired during the phylogenetic evolution.

The law-like process of complexity increase
The emergence of increasingly complex organisms during evolution is the product of an ever-repeating, law-like process of complexity increase. With each round, this process causes the acquisition of a new key innovation, which results in the emergence of a new phylogenetic cluster. At the end of each round, the corre-

sponding generation sequences own an on average higher probability of multiplication compared to their phylogenetic ancestors without key innovation and adapted to the same kind of niche. Hence, the complexity increase occurs in ever-repeating steps, which add successively to the tree of life new phylogenetic clusters of key innovation-owning forms. The individual generation sequence of the forms at the top of the phylogenetic tree owns therefore the relative highest likelihood to continue over time.

Life

Life is the incessantly ongoing multiplication of reproductively interacting molecular forms that consist of a limited number of building blocks. On earth these are DNA, RNA, proteins and biological membranes. It starts with spontaneously emerging molecular interaction communities of relatively simple corresponding precursor forms. Due to exhaustible resources and the need for sustainability, the multiplication of these very first multiplying forms becomes subjected to the lawlike process of complexity increase. This corresponds to the starting point of the generation sequences of life, which can continue over deep time due to the stepwise emergence of multiplying forms with a higher average probability of multiplication and thereby ensuring the ongoing multiplication of the underlying molecular interaction community.

Notes

The main objective of the notes is to provide references for the many facts mentioned in the text. I'm aware that many more could have been given and I apologize to those authors who are not listed but deserved to be mentioned.

Bennett C. H. (2003): How to define complexity in physics and why. In *From Complexity to Life*, ed. Gregersen N.H:, pp. 34-46. New York: Oxford University Press.

Bennett P. M. (2008): Plasmid encoded antibiotic resistance: acquisition and transfer of antibiotic resistance genes in bacteria. Br J Pharmacol. Mar; 153(Suppl 1), pp. S347-S357

Bonner J.T. (2006): Why Size Matters – from bacteria to blue whales. Princeton NJ: Princeton University Press

Bonner J.T. (2017): Sixty Years of Biology - Essays on Evolution and Development. Princeton NJ: Princeton University Press

Byrne R.W. (1994): The evolution of intelligence. In *Behaviour and Evolution*, ed. Slater P.J.B. & Halliday T.R., pp. 223-265. Cambridge: University Press.

Campbell N.A. et al. (2008): Biology, 8th Edition. New York: Pearson Benjamin Cummings.

Carroll S.B. (2001): Chance and necessity: the evolution of morphological complexity and diversity. Nature 409 (6823), pp. 1102-9.

Clokie M. R.J., Millard A. D., Letarov A. V. and Heaphy S. (2011): Phages in nature. Bacteriophage 1:1, pp. 31-45

Conway Morris S. (2003): Life's solution. Inevitable Humans in a Lonely Universe. New York: Cambridge University Press.

Cooper S. and Helmstetter C.E. (1968). Chromosome replication and the division cycle of Escherichia coli B/r. Journal of Molecular Biology. 31 (3), pp. 519-40

Cooper S. (2012): Bacterial Growth and Division. Amsterdam: Elsevier Academic Press.

Courtillot V. (1999): Evolutionary Catastrophes: The Science of Mass Extinction. Cambridge: Cambridge University Press

Darwin C. (1859): On the origin of species. London: John Murray.

Davies P. (1998); The fifth miracle. London: Penguin Books.

Davies P. (1999): The origin of life. London: Penguin Books.

Davies P. (2003): Complexity and the arrow of time. In *From Complexity to Life*, ed. Gregersen N.H:, pp. 72-92. New York: Oxford University Press.

Dawkins R. (1976): The Selfish Gene. Oxford UK: Oxford University Press.

De Duve C. (1995): Vital Dust. New York: Basic Books.

De Meeûs T., Prugnolle F., Agnew P. (2007): Asexual reproduction: genetics and evolutionary aspects. Cell Mol Life Sci. 64(11), pp. 1355-72

Doolittle W.F. (2000): Uprooting the tree of life. Scientific American 282 (2), pp. 90-95

Eigen M. (1971): Self-organization of matter and the evolution of biological macromolecules. Naturwissenschaften 58, pp. 465-523

Eldredge N. (1995): Reinventing Darwin. New York: John Wiley & Sons, Inc.

Embley T. M. and Martin W. (2006): Eukaryotic evolution, changes and challenges. Nature volume 440, pp. 623-630

Feduccia A. (1996): The origin and evolution of birds. New Haven, Conn., Yale University Press

Fontana W. and Buss L.W. (1994): The arrival of the fittest: Towards a theory of biological organization. Bulletin of Mathematical Biology 56, pp. 1-64

Fox Keller E. (1995): Refiguring Life: Metaphors of Twentieth-century Biology. Irvine. Columbia University Press.

Gee H. (2000): Deep Time. Cladistics, the revolution in evolution. London: Fourth Estate Limited.

Gregersen N.H. (2003): From Complexity to Life. New York: Oxford University Press.

Goodwin B. (1994): How the Leopard Changed Its Spots. New York: Charles Scribner's Sons.

Gould S. J. (2002): The structure of evolutionary theory. Cambridge (MA): Harvard University Press.

Gould S. J. and Eldredge N. (1977). Punctuated equilibria: the tempo and mode of evolution reconsidered. Paleobiology 3 (2), pp. 115-151

Gribbin J. (2004): Deep Simplicity - Chaos, Complexity and the Emergence of Life. London: Penguin Books Ltd.

Haken H. (1994): Das einmalige und das gesetzmässige in Physik und Biologie aus der Sicht der Synergetik. In: *Evolution, Entwicklung und Organisation in der Natur*, ed. Braitenberg V. & Hosp I., pp. 24-34. Hamburg: Rowohlt Verlag.

Hallam A. and Wignall P.B. (1997): Mass extinctions and their aftermath. Oxford: Oxford University Press

Halliday T.R. (1994): Sex and evolution. In *Behaviour and Evolution*, ed. Slater P.J.B. & Halliday T.R., pp. 150-192. Cambridge: University Press.

Helfman, G., Collette, B, Facey, D. (1997): The Diversity of Fishes. Hoboken, New Jersey: Blackwell Publishing.

Heylighen F. (1996): The Growth of Structural and Functional Complexity during Evolution. In *The Evolution of Complexity*, ed. F. Heylighen & D. Aerts, pp. 17-44. Kluwer Academic Publishers, Dordrecht NL

Hurst L.D. and Hamilton W.D. (1992): Cytoplasmic Fusion and the Nature of Sexes. Proceedings: Biological Sciences, Vol. 247, No. 1320, pp. 189-194

Jurkevitch E. (2007): Predatory Behaviors in Bacteria—Diversity and Transitions. Microbe, Volume 2, Number 2, pp. 67-73

Kauffman S. (1995a): At home in the universe: The search for laws of self-organization and complexity. New York: Oxford University Press

Kauffman S. (1995b): What is life?: was Schrödinger right?. In *What is Life? The next fifty years: Speculations on the future of biology*, ed. Murphy M.P. and O'Neill L.A.J., pp. 83-114. Cambridge UK: Cambridge University Press

Kauffman S. (2003): The emergence of autonomous agents. In *From Complexity to Life*, ed. Gregersen N.H., pp. 47-71. New York: Oxford University Press

Knoll A.H. (2003): Life on a Young Planet; the first three billion years of evolution on earth. Princeton New Jersey: Princeton University Press

Lenton T. & Watson A. (2011): Revolutions That Made The Earth. New York: Oxford University Press.

Lloyd E.A. (1994): The structure and confirmation of evolutionary theory. Princeton New Jersey: Princeton University Press

Margulis L. (1993): Symbiosis in Cell Evolution. 2nd edition. New York: W.H. Freeman and Company

Magulis L. (1998): The symbiotic planet. A new look at evolution. London: Weidenfeld & Nicolson

Martin W. and Müller M. (1998): The hydrogen hypothesis for the first eukaryote. Nature volume 392, pp. 37-41

Maynard Smith J. and Szathmáry E. (1995): The Major Transitions in Evolution. Oxford, UK: Oxford University Press.

Miller G. (2001): The Mating Mind. New York: Anchor Books.

Miller S.L. (1953): A production of amino acids under possible primitive earth conditions. Science, New Series, Volume 117, 3046, pp: 528-29

Mitchell M. (2009): Complexity a guided tour. Oxford UK: Oxford University Press

Mora C., Tittensor D.P., Adl S., Simpson A.G.B., Worm B. (2011): How Many Species Are There on Earth and in the Ocean? PLoS Biol 9(8): e1001127

Murphy M. P. & O'Neill L.A.J. (1995): What is life? The next fifty years. Speculations on the future of biology. Cambridge: University Press

Ochman H., Davalos L.M. (2006): The nature and dynamics of bacterial genomes. Science 311(5768) pp. 1730-3

Peretó J. (2005): Controversies on the origin of life. International Microbiology 8, pp. 23-31

Pianka E. R. (2000): Evolutionary Ecology. Sixth Edition. San Francisco: Benjamin-Cummings, Addison-Wesley-Longman.

Rose S. (1997): Lifelines. London: Penguin Press

Rothenberg D. (2012): Survival of the Beautiful - Art, Science and Evolution. London: Bloomsbury Publishing

Ruf M.J. (2003): Die Interaktionstheorie: Steigende Komplexität durch Vermehrungsprozesse. Würzburg: Deutscher Wissenschafts-Verlag.

Schneider E.D. and Kay J.J. (1995): Order from disorder: the thermodynamics of complexity in biology. In What is life? The next fifty years, ed. Murphy M.P. and O'Neill A.J., pp.161-174. Cambridge: University Press

Schrödinger E. (1944): What is life? Cambridge UK: Cambridge University Press

Shapiro R. (2007): A simpler origin for life. Scientific American 296, pp. 46-53

Soltis D.E., Bell C.D., Kim S., Soltis P.S. (2008): Origin and Early Evolution of Angiosperms. Ann. N.Y. Acad. Sci. 1133, pp. 3-25

Spiegelman S. (1970): Extracellular evolution of replicating molecules. In The Neuro Sciences: A Second Study Program, ed. Schmitt F.O., pp. 927-945. New York: Rockefeller University Press

Stearns S.C. and Hoekstra R.F. (2005): Evolution an introduction. 2nd edition. New York: Oxford University Press

Stewart I. (1998): Life's other secret - the new mathematics of the living world. Cambridge UK: Cambridge University Press

Stewart I. (2003): The second law of gravitics and the fours law of thermodynamics. In From Complexity to Life, ed. Gregersen N.H., pp. 114 -150. New York: Oxford University Press

Tattersall I. (1998): Becoming human - evolution and human uniqueness. New York: Oxford University Press

The Tree of Life web project (ToL) website: http://tolweb.org/tree/phylogeny.html

Van Valen L. (1973): A new evolutionary law. Evolutionary Theory. 1, pp. 1-30.

Vrba E.S. (1984): Evolutionary pattern and process in the sister-group Alcelaphini-Aepycerotini. In *Living Fossils*, ed. Eldredge N. and Stanley S.M., pp. 62-79. New York: Springer-Verlag

Wächtershäuser G. (2000): Origin of Life: Life as We Don't Know It. Science 289, pp. 1307-08

Watson H. (2015): Biological membranes. Essays Biochem.59, pp. 43-69

Wentworth P. Jr, Janda K.D. (2001): Catalytic antibodies: structure and function. Cell Biochem Biophys; 35(1), pp. 63-87

Wesson R. (1991): Beyond Natural Selection. Cambridge/Mass.: MIT Press

White D. (2007): The Physiology and Biochemistry of Prokaryotes (3rd Ed). New York: Oxford University Press

Williams G.C. (1966): Adaptation and Natural Selection. Princeton NJ: Princeton University Press

Wilson E. O. (2010): The Diversity of Life. Cambridge MA: Harvard University Press

Wing S.L. and Boucher L.D. (1998): Ecological Aspects Of The Cretaceous Flowering Plant Radiation. Annual Review of Earth and Planetary Sciences. 26 (1), pp. 379-421

Wynne-Edwards V.C. (1962): Animal dispersion in relation to social behaviour. Edinburgh: Oliver & Boyd

Wynne-Edwards V.C. (1965): Self-regulating systems in populations of animals. Science 147, pp. 1543-48

Yap M.L. and Rossmann M.G. (2014): Structure and function of bacteriophage T4. Future Microbiology. October 2014; 9, pp. 1319-27

www.ingramcontent.com/pod-product-compliance
Lightning Source LLC
Chambersburg PA
CBHW021558210326
41599CB00010B/505